Lecture Notes in Physics

Founding Editors
Wolf Beiglböck
Jürgen Ehlers
Klaus Hepp
Hans-Arwed Weidenmüller

Volume 1038

Series Editors
Roberta Citro, Salerno, Italy
Peter Hänggi, Augsburg, Germany
Betti Hartmann, London, UK
Morten Hjorth-Jensen, Oslo, Norway
Maciej Lewenstein, Barcelona, Spain
Satya N. Majumdar, Orsay, France
Luciano Rezzolla, Frankfurt am Main, Germany
Angel Rubio, Hamburg, Germany
Wolfgang Schleich, Ulm, Germany
Stefan Theisen, Potsdam, Germany
James D. Wells, Ann Arbor, MI, USA
Gary P. Zank, Huntsville, AL, USA

The series Lecture Notes in Physics (LNP), founded in 1969, reports new developments in physics research and teaching - quickly and informally, but with a high quality and the explicit aim to summarize and communicate current knowledge in an accessible way. Books published in this series are conceived as bridging material between advanced graduate textbooks and the forefront of research and to serve three purposes:

- to be a compact and modern up-to-date source of reference on a well-defined topic;
- to serve as an accessible introduction to the field to postgraduate students and non-specialist researchers from related areas;
- to be a source of advanced teaching material for specialized seminars, courses and schools.

Both monographs and multi-author volumes will be considered for publication. Edited volumes should however consist of a very limited number of contributions only. Proceedings will not be considered for LNP.

Volumes published in LNP are disseminated both in print and in electronic formats, the electronic archive being available at springerlink.com. The series content is indexed, abstracted and referenced by many abstracting and information services, bibliographic networks, subscription agencies, library networks, and consortia.

Proposals should be sent to a member of the Editorial Board, or directly to the responsible editor at Springer:

Dr Lisa Scalone
lisa.scalone@springernature.com

Stefano Olivares

A Student's Guide to Quantum Computing

 Springer

Stefano Olivares
Physics Dept.
University of Milan
Milan, Italy

ISSN 0075-8450 ISSN 1616-6361 (electronic)
Lecture Notes in Physics
ISBN 978-3-031-83360-1 ISBN 978-3-031-83361-8 (eBook)
https://doi.org/10.1007/978-3-031-83361-8

© The Editor(s) (if applicable) and The Author(s), under exclusive license to Springer Nature Switzerland AG 2025

This work is subject to copyright. All rights are solely and exclusively licensed by the Publisher, whether the whole or part of the material is concerned, specifically the rights of translation, reprinting, reuse of illustrations, recitation, broadcasting, reproduction on microfilms or in any other physical way, and transmission or information storage and retrieval, electronic adaptation, computer software, or by similar or dissimilar methodology now known or hereafter developed.
The use of general descriptive names, registered names, trademarks, service marks, etc. in this publication does not imply, even in the absence of a specific statement, that such names are exempt from the relevant protective laws and regulations and therefore free for general use.
The publisher, the authors and the editors are safe to assume that the advice and information in this book are believed to be true and accurate at the date of publication. Neither the publisher nor the authors or the editors give a warranty, expressed or implied, with respect to the material contained herein or for any errors or omissions that may have been made. The publisher remains neutral with regard to jurisdictional claims in published maps and institutional affiliations.

This Springer imprint is published by the registered company Springer Nature Switzerland AG
The registered company address is: Gewerbestrasse 11, 6330 Cham, Switzerland

If disposing of this product, please recycle the paper.

To my family

Preface

> *"Quantum computation is a new conceptual arena for trying to come to a better understanding of quantum weirdness."*
>
> — N. D. Mermin

"At eighty-nine I cannot be productive scientifically; what remains to me is the possibility of following the advances my work prepared and of responding to the wishes of people struggling for truth and knowledge, especially young people, by repeating my lectures here and there."[1]

These words by Max Planck echoed in my mind when I began teaching Quantum Optics and Quantum Computing at the University of Milan. My goal was not only to impart the fundamental principles and tools of these specific fields of research but also, and more importantly, to highlight how quantum theory, laid out by Planck's work, has far-reaching implications in scenarios that physics students may find difficult to imagine.

Reading this guide, the reader will find not only chapters dedicated to the basic theoretical aspects of quantum computation—such as quantum gates, quantum algorithms, and quantum error correction—but also chapters covering fundamental concepts of quantum optics and cavity quantum electrodynamics. These topics are essential for understanding the operating principles behind the most well-known experimental implementations of qubits and quantum computation itself.

Indeed, there are many books for beginners as for advanced readers on the subject of quantum information and, in particular, quantum computation. The student or the researcher can find the one he/she prefers according to his/her own interests, ranging from the quantum algorithms to the physical implementations of quantum information processing and computation. These pages are intended for undergraduate students with a basic understanding of quantum mechanics, as well as researchers interested in the fundamental aspects of quantum computation and the physical principles behind its primary implementations.

As a "student's guide," the book aims to be self-contained. It begins with a brief review of classical logic concepts and a concise introduction to quantum mechanics.

---

[1] J. L. Heilbron, *The dilemmas of an upright man* (University of California Press, 1986).

All the necessary elements for understanding the material are provided to the reader. Additionally, experts can find references to more specialized textbooks and research papers at the end of each chapter for deeper exploration of the topics.

Nevertheless, this book includes numerous detailed explanations and explicit calculations, reflecting over a decade of continuous dialogue with the students attending my lectures. Throughout these years, I have strived to address their questions and clarify their doubts. This extensive interaction has shaped the content, ensuring it meets the needs of those seeking a deeper understanding of the subject matter, while still maintaining the context of an introductory course.

I have chosen to reference only a few key papers, as the reader will notice, often opting for seminal works rather than the most recent publications. This decision reflects the fact that this book is not intended to be a comprehensive manual on quantum computing. Given the rapidly expanding bibliography resulting from recent advances, it is essential to provide a concise exposition and useful tools. This approach helps both the students and the advanced readers grasp the fundamental aspects of this topic.

Only few, selected problems are proposed to bring the reader to a better understanding of the subject and of the mathematical tools, but avoiding any overload of work. The solutions of the problems marked with the symbol " ♣ " can be found at the end of the book.

This is how the material is organized. The basic elements of classical logic are presented in Chap. 1, where we also introduce the Pauli matrices and a non-classical gate: the Hadamard transformation. Chapter 2 covers the postulates of quantum mechanics, with particular emphasis on two-level systems and their dynamics. It also includes discussions on the density operator and entanglement measures.

The first meeting with quantum computation occurs in Chap. 3, where we explore how quantum mechanics can be leveraged for computational tasks. This chapter presents the fundamental quantum logic gates and their role in quantum algorithms, using the circuit representation of quantum computation. The universality of single-qubit and controlled not (CNOT) gates is also discussed and a universal set of quantum gates is introduced. In this chapter, the requirements to have a quantum advantage are given as well (Gottesman–Knill theorem).

Chapter 4 describes the basic aspects of the classical and quantum deterministic Turing machines and universal computers. Moreover, we offer a brief overview of the main complexity classes, providing a first glance at the "complexity zoo."

Chapter 5 illustrates the quantum Fourier transform and its application to Shor's factoring algorithm, while Chap. 6 explains Grover's search algorithm and describes the quantum search on a complete graph via a continuous-time quantum walk.

Chapters 7 and 8 address the crucial issues of noise and errors in quantum computation. The effect of noise is studied using the quantum operation approach, while quantum error correction is demonstrated with the simple but relevant example of the three-qubit code, considering bit-flip, phase-flip, and bit-phase-flip errors. The reader can find some details about the fault-tolerant quantum computation together with the threshold theorem for quantum computation.

The final part of this guide is dedicated to the main physical implementations of quantum computation, and, in particular, to the physical realizations of qubits. Chapter 9 presents the fundamental aspects of two-level systems, such as spin–1/2 particles and two-level atoms, along with the basics of cavity quantum electrodynamics (the Rabi model and the Jaynes–Cummings model). This framework is very common and can be applied in different physical scenarios. The chapter contains a section dedicated to photonic qubits, that not only are at the basis of optical quantum computing, but they find application in other contexts, such as in boson sampling and in quantum communication.

Computation using trapped ions is discussed in Chap. 10, and the physics of superconducting qubits in the charge and transmon regimes is outlined in Chap. 11. Finally, Chap. 12 explores an approach based on the adiabatic evolution of a quantum system to address computational problems and its application to the integer number factoring problem is provided.

As I was finishing this work, I realized that I would have liked to add more aspects regarding the fascinating topic of quantum computing, but I would have risked making it too specialized, going beyond its initial goal: a student's guide that allows the reader to take the first steps (and not only) in the "conceptual arena" that is quantum computing!

Milan, Italy
December 2024

Stefano Olivares

Acknowledgments

First and foremost, I would like to acknowledge the invaluable suggestions and comments I received from my students. Their feedback has been instrumental in improving the content and exposition of these subjects year after year. I am deeply grateful for their contributions, which have significantly enhanced the quality of this work. I also wish to express my heartfelt gratitude to my colleagues at the University of Milan. In particular, I would like to thank Matteo G. A. Paris, Dario Tamascelli, Alessandro Ferraro, Marco G. Genoni, Claudia Benedetti, Simone Cialdi, and Paolo Arosio for their insightful discussions. Their expertise and encouragement have been essential in the development of this book.

Contents

1 Basic Concepts of Classical Logic 1
  1.1 Abstract Representation of Bits 1
  1.2 Classical Logical Operations 3
      1.2.1 Reversible Logical Operations and Permutations 4
  1.3 Single-Bit Reversible Operations 5
  1.4 Two-Bit Reversible Operations 5
      1.4.1 SWAP 5
      1.4.2 Controlled NOT 6
      1.4.3 SWAP, CNOT and Pauli Matrices 8
      1.4.4 The Hadamard Transformation 9
  Problems 9
  Further Readings 10

2 Elements of Quantum Mechanics 11
  2.1 Dirac Notation (in Brief) 11
  2.2 Quantum Bits: Qubits 13
      2.2.1 The Bloch Sphere 14
      2.2.2 Multiple Qubit States 15
  2.3 Postulates of Quantum Mechanics 15
  2.4 Quantum Two-Level System: Explicit Analysis 16
  2.5 Structure of 1-Qubit Unitary Transformations 18
      2.5.1 Linear Transformations and Pauli Matrices 19
  2.6 Quantum States, Density Operator and Density Matrix 20
      2.6.1 Pure States and Statistical Mixtures 21
      2.6.2 Density Operator of a Single Qubit 22
  2.7 The Partial Trace 22
      2.7.1 Purification of Mixed Quantum States 24
      2.7.2 Conditional States 24
  2.8 Entanglement of Two-Qubit States 25
      2.8.1 Entropy of Entanglement 25
      2.8.2 Concurrence 27
  2.9 Quantum Measurements and POVMs 28
  Problems 29
  Further Readings 30

3 Quantum Mechanics as Computation ... 31
- 3.1 Quantum Logic Gates ... 31
  - 3.1.1 Single Qubit Gates ... 32
  - 3.1.2 Single Qubit Gates and Bloch Sphere Rotations ... 33
  - 3.1.3 Two-Qubit Gates: The CNOT Gate ... 34
- 3.2 Measurement on Qubits ... 36
- 3.3 Applications and Examples ... 36
  - 3.3.1 CNOT and No-Cloning Theorem ... 36
  - 3.3.2 Bell States and Bell Measurement ... 37
  - 3.3.3 Quantum Teleportation ... 38
- 3.4 The Standard Computational Process ... 40
  - 3.4.1 Realistic Computation ... 40
- 3.5 Circuit Identities ... 41
- 3.6 Introduction to Quantum Algorithms ... 42
  - 3.6.1 Deutsch Algorithm ... 43
  - 3.6.2 Deutsch–Jozsa Algorithm ... 45
  - 3.6.3 Bernstein–Vazirani Algorithm ... 47
- 3.7 Classical Logic with Quantum Computers ... 49
  - 3.7.1 The Toffoli Gate ... 49
  - 3.7.2 The Fredkin Gate ... 51
- 3.8 Universal Quantum Gates ... 52
  - 3.8.1 Universality of Two-Level Unitaries ... 52
  - 3.8.2 Universality of Single-Qubit and CNOT Gates ... 54
  - 3.8.3 Set of Universal Quantum Gates ... 57
  - 3.8.4 Approximation of Single-Qubit Gates ... 57
- 3.9 Universality and Quantum Advantage ... 60
- Problems ... 61
- Further Readings ... 61

4 Universal Computers and Computational Complexity ... 63
- 4.1 The Turing Machine ... 63
- 4.2 The Quantum Turing Machine ... 64
- 4.3 Important Classical and Quantum Complexity Classes ... 65
- Further Readings ... 68

5 Quantum Fourier Transform and Shor's Factoring Algorithm ... 69
- 5.1 Discrete Fourier Transform and QFT ... 69
- 5.2 The Phase Estimation Protocol ... 74
- 5.3 The Factoring Algorithm (Shor's Algorithm) ... 79
  - 5.3.1 Order-Finding Protocol ... 81
  - 5.3.2 Continued-Fraction Algorithm ... 84
  - 5.3.3 The Factoring Algorithm ... 85
  - 5.3.4 Example: Factorization of the Number 15 ... 86
- 5.4 The RSA Algorithm ... 88
- Problems ... 89
- Further Reading ... 90

6 Quantum Search Algorithm ... 91
- 6.1 Quantum Search as Standard Computational Process ... 91
- 6.2 Quantum Search: The Grover Operator ... 92
  - 6.2.1 Geometric Interpretation of the Grover Operator ... 93
  - 6.2.2 Number of Iterations and Error Probability ... 95
  - 6.2.3 Quantum Counting ... 96
  - 6.2.4 Example of Quantum Search ... 97
- 6.3 Quantum Search and Unitary Evolution ... 98
- 6.4 Grover's Algorithm and Continuous-Time Quantum Walks ... 99
- Problems ... 101
- Further Readings ... 102

7 Quantum Operations ... 103
- 7.1 Environment and Quantum Operations ... 103
- 7.2 Physical Interpretation of Quantum Operations ... 106
- 7.3 The Choi–Jamiołkowski Isomorphism ... 106
- 7.4 Geometric Picture of Single-Qubit Operations ... 110
  - 7.4.1 Bit Flip Operation ... 111
  - 7.4.2 Phase Flip Operation ... 112
  - 7.4.3 Bit-Phase Flip Operation ... 112
  - 7.4.4 Depolarizing Channel ... 113
- 7.5 Amplitude Damping Channel ... 114
- 7.6 Generalized Amplitude Damping Channel ... 116
  - 7.6.1 Approaching the Thermal Equilibrium ... 116
- 7.7 Phase Damping Channel ... 117
- Problems ... 118
- Further Readings ... 118

8 Basics of Quantum Error Correction ... 119
- 8.1 Quantum Error-Correcting Code and Error Correction Conditions ... 119
- 8.2 The Binary Symmetric Channel ... 120
  - 8.2.1 The Three-Bit Code ... 120
- 8.3 Quantum Error Correction: The Three-Qubit Code ... 121
  - 8.3.1 Correction of Bit Flip Error ... 121
  - 8.3.2 Correction of Phase Flip Error ... 125
  - 8.3.3 Correction of Any Error: The Shor Code ... 126
- 8.4 Foult-Tolerant Quantum Computation ... 127
  - 8.4.1 The Threshold Theorem ... 128
- Problems ... 128
- Further Reading ... 129

9 Two-Level Systems and Photonic Qubits ... 131
- 9.1 Universal Computation with Spins ... 131
  - 9.1.1 Interaction Between a Spin–1/2 Particle and a Magnetic Field ... 131

		9.1.2	Spin Qubit and Hadamard Transformation	133

- 9.1.2 Spin Qubit and Hadamard Transformation 133
- 9.1.3 Manipulation of Single Qubit: Nuclear Magnetic Resonance 133
- 9.1.4 How to Realize a CNOT Gate with Spin Systems 135
- 9.1.5 Exchange Interactions and CNOT Gate 136
- 9.1.6 Further Considerations 140
- 9.2 Interaction Between Atoms and Light: Cavity QED 141
  - 9.2.1 Interaction Between a Two-Level Atom and a Classical Electric Field 141
- 9.3 The Quantum Description of Light 144
- 9.4 Photonic Qubits 145
- 9.5 The Jaynes–Cummings Model 146
  - 9.5.1 Vacuum Rabi Oscillations: Quantum Circuit 149
- Problems 151
- Further Readings 151

10 Quantum Computation with Trapped Ions 153
- 10.1 The Linear Paul Trap (in Brief) 153
- 10.2 The Ion Chain 156
- 10.3 Quantum Motion of the Ion Chain 157
- 10.4 Single-Qubit Gates with Trapped Ions 160
- 10.5 CNOT Gate with Trapped Ions 161
- 10.6 Hyperfine and Optical Qubits 164
- Problems 164
- Further Readings 164

11 Superconducting Qubits: Charge and Transmon Qubit 165
- 11.1 The LC Circuit as a Quantum Harmonic Oscillator 165
  - 11.1.1 Quantization of the LC Circuit 166
- 11.2 The Josephson Junction and the SQUID 166
  - 11.2.1 Quantization of the Josephson Junction and SQUID Hamiltonians 170
- 11.3 The Charge Qubit 172
- 11.4 Charge Qubit and Capacitive Coupling with a 1-D Resonator 175
- 11.5 The Transmon Qubit 177
- Problems 180
- Further Readings 181

12 Quantum Computation and Adiabatic Evolution 183
- 12.1 Clauses and Instances of Satisfiability 183
- 12.2 The Adiabatic Theorem 185
- 12.3 Finding the Solutions Through the Adiabatic Evolution 186
- 12.4 One-qubit Example of Adiabatic Quantum Computation 188
- 12.5 Factorization with Adiabatic Evolution 191
- Further Readings 194

A	**Interaction Picture**	195
B	**The Fabry–Perot Cavity**	197

Solutions 203

Index 215

Basic Concepts of Classical Logic

Abstract

In this chapter we give a brief introduction to classical logic introducing the main (classical) logic gates acting on strings of bits. To represent the logical values of the bits, namely "0" and "1", we exploit the same symbols used in the Dirac notations, that is $|0\rangle$ and $|1\rangle$. Focusing on the single- and two-qubit reversible gates we arrive at the representation of the logic gates through the Pauli matrices, that will be useful throughout the rest of this book. The controlled not (CNOT) and the Hadamard gates are introduced as well.

1.1 Abstract Representation of Bits

Classical information is carried by numerical variables and it is extremely useful to use the binary representation $\{0, 1\}$ in order to encode it. If we consider four binary variables $x_k \in \{0, 1\}$, $k = 0, \ldots, 3$, an integer number x can be written in binary notation as follows:

$$x \to x_3\, x_2\, x_1\, x_0,$$
$$= x_3 \times 2^3 + x_2 \times 2^2 + x_1 \times 2^1 + x_0 \times 2^0. \tag{1.1}$$

For instance, $1001 \to 1 \times 2^3 + 0 \times 2^2 + 0 \times 2^1 + 1 \times 2^0 = 9$.

The *amount of information* carried by the binary variable is called *bit*. Each binary variable can take only two values, thus a sequence of n binary variables can be actually used to name $N = 2^n$ different numbers. The length of a string tells us the space required to hold the number. We can consider $\log_2 N = \log_2 2^n = n$ a measure of the information. Note that a single bit carries $\log_2 2 = 1$ bit of information.

© The Author(s), under exclusive license to Springer Nature Switzerland AG 2025
S. Olivares, *A Student's Guide to Quantum Computing*, Lecture Notes
in Physics 1038, https://doi.org/10.1007/978-3-031-83361-8_1

Instead of using the symbols "0" and "1", we will use the *abstract* symbols $|0\rangle$ and $|1\rangle$, respectively. By using this formalism, the binary string "1001" rewrites as:[1]

$$1001 \to |1\rangle|0\rangle|0\rangle|1\rangle, \tag{1.2}$$

which represents the *state* of the four classical bit carrying the information. It is worth noting that, in reality, each symbol $|x\rangle$, $x = 0, 1$, is associated with a *physical* entity. Therefore, we can identify the numerical value of the classical bit with the bit itself. For the sake of simplicity, we can use the following notation:

$$|1001\rangle \equiv |1\rangle|0\rangle|0\rangle|1\rangle \tag{1.3}$$

or also write:

$$|1001\rangle \equiv |9\rangle_4 \tag{1.4}$$

where we used the decimal notation "9" to represent the binary value "1001" and the subscript "4" refers to the four bits we used to encode the number (indeed, mathematically, the two binary strings "1001" and "0000001001" represent the same digital number "9", but, physically, the first involves only four bits, the second employs ten bits!!).

It is possible to associate two column vectors with $|0\rangle$ and $|1\rangle$ as follows:

$$|0\rangle \to \begin{pmatrix} 1 \\ 0 \end{pmatrix}, \quad \text{and} \quad |1\rangle \to \begin{pmatrix} 0 \\ 1 \end{pmatrix}. \tag{1.5}$$

We clearly see that the two vectors are orthonormal. Now, we note that the symbol $|1\rangle|0\rangle|0\rangle|1\rangle$ is a short-hand for the tensor product of four single-bit 2-dimensional vector, namely:

$$|1\rangle|0\rangle|0\rangle|1\rangle \equiv |1\rangle \otimes |0\rangle \otimes |0\rangle \otimes |1\rangle. \tag{1.6}$$

Let's focus on a 4-dimensional space, with orthonormal basis:

$$|0\rangle_2 = |00\rangle \to \begin{pmatrix} 1 \\ 0 \\ 0 \\ 0 \end{pmatrix}, \quad |1\rangle_2 = |01\rangle \to \begin{pmatrix} 0 \\ 1 \\ 0 \\ 0 \end{pmatrix}, \tag{1.7a}$$

---

[1] We will se later on the mathematical framework of this formalism.

$$|2\rangle_2 = |10\rangle \to \begin{pmatrix} 0 \\ 0 \\ 1 \\ 0 \end{pmatrix}, \quad |3\rangle_2 = |11\rangle \to \begin{pmatrix} 0 \\ 0 \\ 0 \\ 1 \end{pmatrix}, \qquad (1.7b)$$

where we explicitly evaluated the tensor product.[2] In this way it is possible to obtain the 2^n-dimensional column vector representing any of the 2^n possible states of n bits.

If $x = (x_0, x_1, \ldots, x_{n-1})^\mathsf{T}$, $x_k \in \{0, 1\}$, $k = 0, \ldots, n-1$, is a column vector associated with the binary representation of an integer $0 \le x < 2^n$, then:

$$x = \sum_{k=0}^{n-1} x_k 2^k, \qquad (1.8)$$

and we have:[3]

$$|x\rangle_n = |x_{n-1}\rangle \otimes \cdots \otimes |x_0\rangle = |x_{n-1} \cdots x_1 x_0\rangle, \qquad (1.9)$$

i.e., $|x\rangle_n$ is the tensor product of the single-bit states $|x_k\rangle$.

1.2 Classical Logical Operations

Any logical or arithmetical operation can be obtained by the composition of three elementary logical operations: "NOT", "AND" and "OR". The NOT operation acts on a single bit, while AND and OR are two-bit operations. Their actions are summarized in the truth tables (Tables 1.1, 1.2, and 1.3).

It is worth noting that the three logical operations introduced above are not independent: given NOT and OR it is possible to obtain the operation AND; analogously, given NOT and AND it is possible to obtain the operation OR. Thus, we can introduce the two *universal* operators "NOR" (NOT OR) and "NAND" (NOT AND):

$$\mathrm{NOR}|x\rangle|y\rangle \equiv |\overline{x \vee y}\rangle = |\overline{x} \wedge \overline{y}\rangle, \qquad (1.10a)$$

$$\mathrm{NAND}|x\rangle|y\rangle \equiv |\overline{x \wedge y}\rangle = |\overline{x} \vee \overline{y}\rangle. \qquad (1.10b)$$

---

[2] The tensor product of the two column vectors $(a_1, \ldots, a_N)^\mathsf{T}$ and $(b_1, \ldots, b_M)^\mathsf{T}$ is a NM-component vector with components indexed by all the MN possible pairs of indices (ν, μ), whose (ν, μ)th component is just the product $a_\nu b_\mu$.

[3] Note that the binary expansion of the column vector $x = (x_0, x_1, \ldots, x_{n-1})^\mathsf{T}$ is $x \to x_{n-1} \cdots x_1 x_0$.

Table 1.1 NOT operation. We used the alternative notation $\text{NOT}|x\rangle = |\bar{x}\rangle$

$\|x\rangle$	$\|\bar{x}\rangle$
$\|0\rangle$	$\|1\rangle$
$\|1\rangle$	$\|0\rangle$

Table 1.2 AND operation. We used the alternative notation $\text{AND}|x\rangle|y\rangle = |x \wedge y\rangle$

$\|x\rangle\|y\rangle$	$\|x \wedge y\rangle$
$\|0\rangle\|0\rangle$	$\|0\rangle$
$\|0\rangle\|1\rangle$	$\|0\rangle$
$\|1\rangle\|0\rangle$	$\|0\rangle$
$\|1\rangle\|1\rangle$	$\|1\rangle$

Table 1.3 OR operation. We used the alternative notation $\text{OR}|x\rangle|y\rangle = |x \vee y\rangle$

$\|x\rangle\|y\rangle$	$\|x \vee y\rangle$
$\|0\rangle\|0\rangle$	$\|0\rangle$
$\|0\rangle\|1\rangle$	$\|1\rangle$
$\|1\rangle\|0\rangle$	$\|1\rangle$
$\|1\rangle\|1\rangle$	$\|1\rangle$

Table 1.4 XOR operation. We used the alternative notation $\text{XOR}|x\rangle|y\rangle = |x \oplus y\rangle$

$\|x\rangle\|y\rangle$	$\|x \oplus y\rangle$
$\|0\rangle\|0\rangle$	$\|0\rangle$
$\|0\rangle\|1\rangle$	$\|1\rangle$
$\|1\rangle\|0\rangle$	$\|1\rangle$
$\|1\rangle\|1\rangle$	$\|0\rangle$

Another useful operator is the XOR, or *exclusive* OR operator, which corresponds to the modulo-2 sum. Its action is summarized in Table 1.4. Note that $|\bar{x}\rangle = |x \oplus 1\rangle$. As a matter of fact the XOR can be reduced to more elementary operations as:

$$|x \oplus y\rangle = \left|(x \vee y) \wedge \overline{(x \wedge y)}\right\rangle. \qquad (1.11)$$

1.2.1 Reversible Logical Operations and Permutations

A logical function is reversible if each output arises from a unique input: it is possible to show that a reversible function should be a *permutation* of the input bit states. The inspection of Tables 1.1, 1.2, 1.3, and 1.4 shows that among the presented operations, only NOT is reversible. Reversibility plays a relevant role in quantum computation, since, as we will see, the general computational process can be modeled with a unitary operation that is indeed reversible.

1.3 Single-Bit Reversible Operations

The NOT is the only reversible (classical) operation acting on single bits (excluding the identity operator $\hat{\mathbb{I}}$, which is a trivial operation). By using the matrix formalism, we can represent NOT with the 2×2 matrix:

$$\mathbf{X} \to \begin{pmatrix} 0 & 1 \\ 1 & 0 \end{pmatrix}. \tag{1.12}$$

Since $\mathbf{X}^2 = \hat{\mathbb{I}} \to \mathbb{1}_2 = \text{diag}(1,1)$ is the 2×2 identity matrix, it follows that $\mathbf{X}$ is invertible and $\mathbf{X} = \mathbf{X}^{-1}$.

It is also instructive to introduce the operators $\mathbf{N}$, the number operator, and $\overline{\mathbf{N}} = \hat{\mathbb{I}} - \mathbf{N}$:

$$\mathbf{N}|x\rangle = x|x\rangle, \quad \text{and} \quad \overline{\mathbf{N}}|x\rangle = \overline{x}|x\rangle, \quad x \in \{0,1\}. \tag{1.13}$$

The corresponding matrices are:

$$\mathbf{N} \to \begin{pmatrix} 0 & 0 \\ 0 & 1 \end{pmatrix}, \quad \text{and} \quad \overline{\mathbf{N}} \to \begin{pmatrix} 1 & 0 \\ 0 & 0 \end{pmatrix}. \tag{1.14}$$

Classically, $\mathbf{N}$ and $\overline{\mathbf{N}}$ are just mathematical operators and do not correspond to a physical operation, e.g. we cannot imagine the meaning of multiplying by 0 the *state*—not the *numerical value*—of a bit... However, they could be useful from the formal point of view.

1.4 Two-Bit Reversible Operations

1.4.1 SWAP

The SWAP operation exchanges the *values* x and y of the two bits $|x\rangle|y\rangle$:

$$\mathbf{S}|x\rangle|y\rangle = |y\rangle|x\rangle. \tag{1.15}$$

If we consider the n-bit state $|x\rangle_n$, then we can define the operator $\mathbf{S}_{hk}$ which acts on the bits h and k, namely:

$$\mathbf{S}_{hk}|x\rangle_n = \mathbf{S}_{hk}|x_{n-1}\rangle \cdots |x_h\rangle \cdots |x_k\rangle \cdots |x_0\rangle,$$
$$= |x_{n-1}\rangle \cdots |x_k\rangle \cdots |x_h\rangle \cdots |x_0\rangle. \tag{1.16}$$

Table 1.5 CNOT operation

$\|x\rangle\|y\rangle$	$\mathbf{C}_{10}$	$\mathbf{C}_{01}$
$\|0\rangle\|0\rangle$	$\|0\rangle\|0\rangle$	$\|0\rangle\|0\rangle$
$\|0\rangle\|1\rangle$	$\|0\rangle\|1\rangle$	$\|1\rangle\|1\rangle$
$\|1\rangle\|0\rangle$	$\|1\rangle\|1\rangle$	$\|1\rangle\|0\rangle$
$\|1\rangle\|1\rangle$	$\|1\rangle\|0\rangle$	$\|0\rangle\|1\rangle$

Since $\mathbf{S}_{hk}\mathbf{S}_{hk} = \hat{\mathbb{I}}$, the SWAP is indeed unitary. It is also possible to represent the SWAP as follows:

$$\mathbf{S}_{hk} = \mathbf{N}_h \otimes \mathbf{N}_k + \overline{\mathbf{N}}_h \otimes \overline{\mathbf{N}}_k + (\mathbf{X}_h \otimes \mathbf{X}_k)\left(\mathbf{N}_h \otimes \overline{\mathbf{N}}_k + \overline{\mathbf{N}}_h \otimes \mathbf{N}_k\right), \quad (1.17)$$

where $\mathbf{N}_k$, $\overline{\mathbf{N}}_k$ and $\mathbf{X}_k$ have been introduced in Sect. 1.3 and act on the k-th bits. Sometimes, we will drop the tensor product symbol and we will write:

$$\mathbf{S}_{hk} = \mathbf{N}_h \mathbf{N}_k + \overline{\mathbf{N}}_h \overline{\mathbf{N}}_k + \mathbf{X}_h \mathbf{X}_k \left(\mathbf{N}_h \overline{\mathbf{N}}_k + \overline{\mathbf{N}}_h \mathbf{N}_k\right), \quad (1.18)$$

The reader can verify the action of the left-hand-side member of Eq. (1.17) by exploiting the properties of the tensor product and recalling that, given two operators $\mathbf{A}_h$ and $\mathbf{B}_k$, acting on the h-th and k-th bits, respectively, one has:

(i) $\mathbf{A}_h \otimes \mathbf{B}_k |x_h\rangle \otimes |x_k\rangle = \mathbf{A}_h |x_h\rangle \otimes \mathbf{B}_k |x_k\rangle$;

(ii) $(\mathbf{A}_h \otimes \mathbf{B}_k)(\mathbf{C}_h \otimes \mathbf{D}_k) = (\mathbf{A}_h \mathbf{C}_h) \otimes (\mathbf{B}_k \mathbf{D}_k)$.

The matrix representation of $\mathbf{S}_{hk}$ is just a single permutation matrix.[4]

1.4.2 Controlled NOT

The controlled-NOT or, in brief, CNOT, is a "workhorse for quantum computation". This operation acts on a *target* bit according to the value of a *control* bit. By definition, $\mathbf{C}_{hk}$ flips the state of the k-th bit (target state) only if the state of the h-th bit (control state) is $|1\rangle$. The action of $\mathbf{C}_{10}$ and $\mathbf{C}_{01}$ is summarized in Table 1.5: we can easily see that they act as permutations on the input basis in which only two elements are exchanged.

---

[4] The explicit form of the permutation matrix associated with $\mathbf{S}_{hk}$ can be obtained starting from the identity matrix and exchanging the h-th and k-th columns.

1.4 Two-Bit Reversible Operations

The matrix representations of $\mathbf{C}_{01}$ and $\mathbf{C}_{10}$ are:

$$\mathbf{C}_{10} \rightarrow \begin{pmatrix} 1 & 0 & 0 & 0 \\ 0 & 1 & 0 & 0 \\ 0 & 0 & 0 & 1 \\ 0 & 0 & 1 & 0 \end{pmatrix}, \quad \mathbf{C}_{01} \rightarrow \begin{pmatrix} 1 & 0 & 0 & 0 \\ 0 & 0 & 0 & 1 \\ 0 & 0 & 1 & 0 \\ 0 & 1 & 0 & 0 \end{pmatrix}, \quad (1.19)$$

respectively.

Note that, in general, we can summarize the action of CNOT as follows:

$$\begin{aligned} \mathbf{C}_{hk}|x\rangle_n &= \mathbf{C}_{hk}|x_{n-1}\rangle \cdots |x_h\rangle \cdots |x_k\rangle \cdots |x_0\rangle, \\ &= |x_{n-1}\rangle \cdots |x_h\rangle \cdots |x_k \oplus x_h\rangle \cdots |x_0\rangle, \end{aligned} \quad (1.20)$$

where we used $|x_k \oplus x_h\rangle = |\overline{x}_k\rangle$ if and only if $|x_h\rangle = |1\rangle$. It is clear that CNOT acts as a generalized XOR.

Now, we introduce the operator:

$$\mathbf{Z} = \overline{\mathbf{N}} - \mathbf{N} \rightarrow \begin{pmatrix} 1 & 0 \\ 0 & -1 \end{pmatrix}, \quad (1.21)$$

and $\mathbf{XZ} = -\mathbf{ZX}$. It is straightforward to see that:

$$\mathbf{Z}|x\rangle = (-1)^x|x\rangle, \quad x \in \{0, 1\}. \quad (1.22)$$

From a classical point of view the action of $\mathbf{Z}$ is meaningless: it multiplies by -1 the state $|1\rangle$—note that the *state* of the bit is multiplied by -1 and not its numerical value!

Since, $\mathbf{N} = \frac{1}{2}(\hat{\mathbb{I}} - \mathbf{Z})$ and $\overline{\mathbf{N}} = \frac{1}{2}(\hat{\mathbb{I}} + \mathbf{Z})$ which directly follows from Eq. (1.22), we can write:[5]

$$\mathbf{C}_{hk} = \frac{1}{2}(\hat{\mathbb{I}} + \mathbf{Z}_h) + \frac{1}{2}(\hat{\mathbb{I}} - \mathbf{Z}_h)\mathbf{X}_k, \quad (1.23a)$$

$$= \frac{1}{2}(\hat{\mathbb{I}} + \mathbf{X}_k) + \frac{1}{2}\mathbf{Z}_h(\hat{\mathbb{I}} - \mathbf{X}_k), \quad (1.23b)$$

where we dropped the tensor product.

---

[5] In order to simplify the formalism, we use the following convention:

$$\mathbf{A}_h \otimes \hat{\mathbb{I}}(|x_h\rangle \otimes |x_k\rangle) \equiv \mathbf{A}_h(|x_h\rangle \otimes |x_k\rangle),$$

i.e., $\mathbf{A}_h \otimes \hat{\mathbb{I}} \equiv \mathbf{A}_h$.

1.4.3 SWAP, CNOT and Pauli Matrices

Substituting Eqs. (1.23) into Eq. (1.33), one find the following interesting identity for the SWAP operator:

$$\mathbf{S}_{hk} = \frac{1}{2}(\hat{\mathbb{I}} + \mathbf{Z}_h \mathbf{Z}_k) + \frac{1}{2}\mathbf{X}_h \mathbf{X}_k (\hat{\mathbb{I}} - \mathbf{Z}_h \mathbf{Z}_k), \tag{1.24}$$

which may be also written as:

$$\mathbf{S}_{hk} = \frac{1}{2}(\hat{\mathbb{I}} + \mathbf{X}_h \mathbf{X}_k - \mathbf{Y}_h \mathbf{Y}_k + \mathbf{Z}_h \mathbf{Z}_k), \tag{1.25}$$

where:[6]

$$\mathbf{Y}_k = \mathbf{Z}_k \mathbf{X}_k \to \begin{pmatrix} 0 & 1 \\ -1 & 0 \end{pmatrix}. \tag{1.26}$$

If, however, we introduce the Pauli operators (and the corresponding 2×2 Pauli matrices):

$$\hat{\sigma}_x \to \sigma_x = \begin{pmatrix} 0 & 1 \\ 1 & 0 \end{pmatrix}, \quad \hat{\sigma}_y \to \sigma_y = \begin{pmatrix} 0 & -i \\ i & 0 \end{pmatrix}, \quad \hat{\sigma}_z \to \sigma_z = \begin{pmatrix} 1 & 0 \\ 0 & -1 \end{pmatrix} \tag{1.27}$$

we have:

$$\mathbf{S}_{hk} = \frac{1}{2}\left(\hat{\mathbb{I}} + \hat{\sigma}_x^{(h)}\hat{\sigma}_x^{(k)} + \hat{\sigma}_y^{(h)}\hat{\sigma}_y^{(k)} + \hat{\sigma}_z^{(h)}\hat{\sigma}_z^{(k)}\right), \tag{1.28}$$

where the superscripts refer to the target qubits.

Analogously we can write the CNOT as:

$$\mathbf{C}_{hk} = \frac{1}{2}\left(\hat{\mathbb{I}} + \hat{\sigma}_x^{(k)} + \hat{\sigma}_z^{(h)} - \hat{\sigma}_z^{(h)}\hat{\sigma}_x^{(k)}\right), \tag{1.29}$$

Pauli matrices, together with the identity matrix, form a basis for the 2×2 matrices and have the following properties:

$$[\hat{\sigma}_x, \hat{\sigma}_y] = \hat{\sigma}_x \hat{\sigma}_y - \hat{\sigma}_y \hat{\sigma}_x = 2i\hat{\sigma}_z, \tag{1.30a}$$

$$[\hat{\sigma}_y, \hat{\sigma}_z] = \hat{\sigma}_y \hat{\sigma}_z - \hat{\sigma}_z \hat{\sigma}_y = 2i\hat{\sigma}_x, \tag{1.30b}$$

---

[6] It is worth noting that in our formalism if $k \neq h$ we have $\mathbf{A}_k \mathbf{B}_h = \mathbf{A}_k \otimes \mathbf{B}_h$, since the two operators refer to different physical entities; the symbol $\mathbf{A}_k \mathbf{B}_k$ represents the composition of the two operators.

$$[\hat{\sigma}_z, \hat{\sigma}_x] = \hat{\sigma}_z\hat{\sigma}_x - \hat{\sigma}_x\hat{\sigma}_z = 2i\hat{\sigma}_y, \qquad (1.30c)$$

or, by introducing the totally antisymmetric tensor ε_{hkl}, $[\hat{\sigma}_h, \hat{\sigma}_k] = 2i\varepsilon_{hkl}\hat{\sigma}_l$.

1.4.4 The Hadamard Transformation

The Hadamard transformation is defined as:

$$\mathbf{H} = \frac{1}{\sqrt{2}}(\mathbf{X} + \mathbf{Z}) \rightarrow \frac{1}{\sqrt{2}}\begin{pmatrix} 1 & 1 \\ 1 & -1 \end{pmatrix}. \qquad (1.31)$$

Though, classically speaking, the action of **H** on $|x\rangle$ is meaningless, since **H** transforms a single-bit state into a linear combination of states, namely:

$$\mathbf{H}|x\rangle = \frac{|0\rangle + (-1)^x|1\rangle}{\sqrt{2}},$$

or, explicitly:

$$\mathbf{H}|0\rangle = \frac{|0\rangle + |1\rangle}{\sqrt{2}}, \quad \text{and} \quad \mathbf{H}|1\rangle = \frac{|0\rangle - |1\rangle}{\sqrt{2}}, \qquad (1.32)$$

this transformation is useful when applied recursively to other operators, as the reader can see from the problems 1.6 and 1.7.

Problems

1.1. Prove that NOR and NAND are universal.

1.2. Verify that $\mathbf{X}|x\rangle = |\bar{x}\rangle$.

1.3. Verify that $\overline{\mathbf{N}}^2 = \overline{\mathbf{N}}$ and $\mathbf{N}\overline{\mathbf{N}} = \overline{\mathbf{N}}\mathbf{N} = \mathbf{0}$.

1.4. Verify that $\mathbf{C}_{hk} = \overline{\mathbf{N}}_h + \mathbf{N}_h\mathbf{X}_k$, where the subscripts refer to the bit affected by the operation.

1.5. ♣ Show that the same action of the SWAP can be obtained by the application of three CNOT operations, namely:

$$\mathbf{S}_{hk} = \mathbf{C}_{hk}\mathbf{C}_{kh}\mathbf{C}_{hk}. \qquad (1.33)$$

1.6. ♣ Show that $\mathbf{HXH} = \mathbf{Z}$ and $\mathbf{HZH} = \mathbf{X}$, that is, the Hadamard transformation allows to transform **X** into **Z** and vice versa.

1.7. ♣ Show that:

$$\mathbf{C}_{hk} = \mathbf{H}_h \mathbf{H}_k \mathbf{C}_{kh} \mathbf{H}_h \mathbf{H}_k, \tag{1.34}$$

where the subscripts have the usual meaning—the Hadamard transformation allows to exchange the roles of the target bit and of the control bit of a CNOT, namely:

$$\mathbf{C}_{hk} \to \mathbf{C}_{kh}.$$

Further Readings

N.D. Mermin, *Quantum Computer Science* (Cambridge University Press, 2007) – Chapter 1
M.A. Nielsen, I.L. Chuang, *Quantum Computation and Quantum Information* (Cambridge University Press, 2010) – Chapter 1

Elements of Quantum Mechanics

2

Abstract

In this chapter we briefly review the theoretical framework of quantum mechanics. In particular, the reader can find the postulates of quantum mechanics and the description of the measurement through the positive operator-valued measures (POVMs). We also introduce the density operator describing the system highlighting the difference between pure and mixed states. The concept of entanglement is also mentioned and quantified through the entropy of the entanglement and the concurrence for two-qubit systems.

2.1 Dirac Notation (in Brief)

Throughout this chapter we use the Dirac bracket notation. An n-dimensional complex vector (or state) is represented with the symbol $|\psi\rangle_n$, that is called "ket". Given two vectors $|\psi\rangle_n$ and $|\phi\rangle_n$, we use the following symbol for the *inner product* (we drop the subscript n): $\langle\psi|(|\phi\rangle) \equiv \langle\psi|\phi\rangle \in \mathbb{C}$. Indeed, $\langle\psi|\phi\rangle$ can be seen as a linear functional associated with the vector $|\psi\rangle$ that takes $|\phi\rangle$ into a complex number. This functional is $(|\psi\rangle)^\dagger = \langle\psi|$, where the symbol $(\cdots)^\dagger$ represents the adjoint operator, and $\langle\psi|$ is called "bra". As usual, the inner product satisfies the following properties:

(i) $\langle\psi|\phi\rangle = \langle\phi|\psi\rangle^*$;

(ii) $\langle\psi|(\alpha|\phi\rangle + \beta|\gamma\rangle) = \alpha\langle\psi|\phi\rangle + \beta\langle\psi|\gamma\rangle, \forall \alpha, \beta \in \mathbb{C}$;

(iii) $\langle\psi|\psi\rangle = 0 \Leftrightarrow |\psi\rangle = 0$.

We can expand the (2^n-dimensional) vector $|\psi\rangle$ as follows:

$$|\psi\rangle = \sum_{x=0}^{2^n-1} \alpha_x |x\rangle, \qquad (2.1)$$

where $\langle x|y\rangle = \delta_{xy}$, δ_{xy} being the Kronecker delta, and $\{|x\rangle\}$ is a basis of the space, that is the Hilbert space, namely, a complex vectorial space with inner product, as we will see in Sect. 2.3. By using the same association between kets and vectors introduced in Sect. 1.1, we have:

$$|\psi\rangle \to \begin{pmatrix} \alpha_0 \\ \alpha_1 \\ \vdots \\ \alpha_{2^n-1} \end{pmatrix}, \quad \text{and} \quad \langle\psi| \to \left(\alpha_0^*, \alpha_1^*, \ldots, \alpha_{2^n-1}^*\right), \qquad (2.2)$$

where $\langle x|\psi\rangle = \alpha_x$ and the basis the vectors $|x\rangle$, $0 \leq x < 2^n$, have been introduced in Sect. 1.1. It is now clear that, with this association, the inner product between bras and kets corresponds to the standard inner product between the corresponding vectors.

Let us now consider the linear operator $\hat{A}$ which acts on a ket $|\psi\rangle$ leading to a new vector, namely $\hat{A}|\psi\rangle = |\psi'\rangle$. We have $(\hat{A}|\psi\rangle)^\dagger = \langle\psi|\hat{A}^\dagger$ and:

$$\langle\phi|\hat{A}|\psi\rangle = \underbrace{\left(\langle\phi|\hat{A}\right)}_{\left(\hat{A}^\dagger|\phi\rangle\right)^\dagger} |\psi\rangle = \langle\phi|\left(\hat{A}|\psi\rangle\right). \qquad (2.3)$$

The *outer product* between $|\psi\rangle$ and $|\phi\rangle$ is an operator $|\psi\rangle\langle\phi|$ whose action on $|\gamma\rangle$ reads:

$$|\psi\rangle\langle\phi|(|\gamma\rangle) = |\psi\rangle\langle\phi|\gamma\rangle \equiv \langle\phi|\gamma\rangle|\psi\rangle. \qquad (2.4)$$

Furthermore, we have:

$$|\psi\rangle\langle\phi| \to \begin{pmatrix} \alpha_0 \\ \alpha_1 \\ \vdots \\ \alpha_{2^n-1} \end{pmatrix} \cdot \left(\beta_0^*, \beta_1^*, \ldots, \beta_{2^n-1}^*\right) \equiv \mathbf{M}, \qquad (2.5)$$

where $\mathbf{M}$ is a $2^n \times 2^n$ matrix with entries $[\mathbf{M}]_{xy} = \alpha_x \beta_y^*$, and we wrote $|\psi\rangle = \sum_x \alpha_x |x\rangle$ and $|\phi\rangle = \sum_y \beta_y |y\rangle$.

The operator $\hat{P}_x = |x\rangle\langle x|$, $0 \le x < 2^n$, is called *projector* onto the vector $|x\rangle$ (indeed, one can define a projector $\hat{P}_\psi = |\psi\rangle\langle\psi|$ onto the state $|\psi\rangle$). Since $\{|x\rangle\}$ is an orthonormal basis for the 2^n-dimensional vector space, we have the following *completeness* relation:

$$\sum_x |x\rangle\langle x| = \hat{\mathbb{I}}, \tag{2.6}$$

that is we have a resolution of the identity operator. The completeness relation may be used to express vectors and operators in a particular orthonormal basis.

2.2 Quantum Bits: Qubits

We consider the complex vector space generated by the two column vectors associated with the bit states $|0\rangle$ and $|1\rangle$ (that is a 2-dimensional complex Hilbert space). Since the two states form a basis for this space, any linear combination, or *superposition*:

$$|\psi\rangle = \alpha|0\rangle + \beta|1\rangle \rightarrow \begin{pmatrix} \alpha \\ \beta \end{pmatrix}, \tag{2.7}$$

where $\alpha, \beta \in \mathbb{C}$, belongs to the space. If $|\alpha|^2 + |\beta|^2 = 1$, i.e., if $|\psi\rangle$ is *normalized*, we will refer to the state (2.7) as *quantum bit* or simply *qubit*. Of course, if $\alpha = 0$ or $\beta = 0$, then $|\psi\rangle = |1\rangle$ or $|\psi\rangle = |0\rangle$, respectively.[1]

The basis $\{|0\rangle, |1\rangle\}$ is called *computational basis* and the information is stored in complex numbers α and β: it follows that in a single qubit it is possible to encode an infinite amount of information. At least potentially … In fact, in order to extract the information we should perform a *measurement* on the qubit: as we will see in the next sections, it is a fundamental aspect of Nature that when we observe a system in the superposition state (2.7), we find it either in the state $|0\rangle$ or in $|1\rangle$ with a probabilities $p(0) = |\alpha|^2$ and $p(0) = |\beta|^2$, that's why $|\alpha|^2 + |\beta|^2 = 1$.[2]

Since $|\alpha|^2 + |\beta|^2 = 1$, we can use the following useful parameterization for the amplitudes of the qubit sate:[3]

$$\alpha = \cos\frac{\theta}{2}, \quad \text{and} \quad \beta = e^{i\phi}\sin\frac{\theta}{2}, \tag{2.8}$$

---

[1] The reader may observe that one should write $|\psi\rangle = e^{i\phi}|1\rangle$ or $|\psi\rangle = e^{i\phi}|0\rangle$, but we will see in Sect. 2.3 that a *global* phase, as $e^{i\phi}$, does not have a physical meaning.

[2] Here we are assuming that the measurement allows to observe as outcomes the state $|0\rangle$ or $|1\rangle$, i.e., the computational basis; of course one may choose a different basis for the measurement, for instance one can also use other computational basis, e.g., $\{|+\rangle, |-\rangle\}$, where $|\pm\rangle = 2^{-1/2}(|0\rangle + |1\rangle)$.

[3] More in general one should have $\alpha = e^{i\delta}\cos\frac{\theta}{2}$ and $\beta = e^{i\phi}\sin\frac{\theta}{2}$, but this is equivalent to add a global phase to the state and, thus, we can set $\delta = 0$.

obtaining:

$$|\psi\rangle = \cos\frac{\theta}{2}|0\rangle + e^{i\phi}\sin\frac{\theta}{2}|1\rangle. \tag{2.9}$$

We will address in Chaps. 9, 10, and 11 some examples of the physical realization of qubits.

2.2.1 The Bloch Sphere

We can associate the following three real numbers with the qubit (2.9):

$$r_x = \sin\theta\,\cos\phi, \quad r_y = \sin\theta\,\sin\phi, \quad r_z = \cos\theta, \tag{2.10}$$

which can be seen as the components of a 3-dimensional vector, i.e.:

$$\boldsymbol{r} = \begin{pmatrix} r_x \\ r_y \\ r_z \end{pmatrix} = \begin{pmatrix} \sin\theta\,\cos\phi \\ \sin\theta\,\sin\phi \\ \cos\theta \end{pmatrix}. \tag{2.11}$$

Furthermore, since $|\boldsymbol{r}| = \sqrt{r_x^2 + r_y^2 + r_z^2} = 1$, $\boldsymbol{r}$ represents a point on the surface of the unit sphere, that is the so-called Bloch sphere. In Fig. 2.1 we show the Bloch sphere and the vectorial representation of a quantum state (the red vector).

In particular we have:

$$|0\rangle \Rightarrow \begin{pmatrix} 0 \\ 0 \\ 1 \end{pmatrix}, \quad \text{and} \quad |1\rangle \Rightarrow \begin{pmatrix} 0 \\ 0 \\ -1 \end{pmatrix}, \tag{2.12}$$

Fig. 2.1 The Bloch sphere is represented by the yellow unit sphere, while the red vector represents a pure state, i.e., a state belonging to the surface of the sphere. We also show the two angles θ (magenta) and ϕ (blue) which identify the quantum state

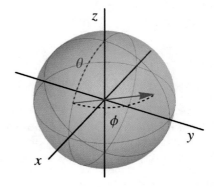

namely, $|0\rangle$ corresponds to the north pole of the Bloch sphere, whereas $|1\rangle$ to its south pole. The state $|\psi\rangle = 2^{-1/2}(|0\rangle + e^{i\phi}|1\rangle)$, with $\phi \in [0, 2\pi)$, corresponds to equatorial states.

2.2.2 Multiple Qubit States

A n-qubit state reads:

$$|\Psi\rangle_n = \sum_{x=0}^{2^n-1} \alpha_x |x\rangle_n, \quad \text{with} \quad \sum_{x=0}^{2^n-1} |\alpha_x|^2 = 1, \tag{2.13}$$

as usual, the subscript n refers to the number of physical entities (qubits) used to encode the information. In particular, the state of two qubits can be written as:

$$|\Psi\rangle_2 = \alpha_{00}|00\rangle + \alpha_{01}|01\rangle + \alpha_{10}|10\rangle + \alpha_{11}|11\rangle, \tag{2.14}$$

with $|\alpha_{00}|^2 + |\alpha_{10}|^2 + |\alpha_{01}|^2 + |\alpha_{11}|^2 = 1$. In this case, each $|\alpha_{xy}|^2$ corresponds to the *joint* probability to find the two qubits of the state (2.14) in the state $|x\,y\rangle$.

2.3 Postulates of Quantum Mechanics

In this section we introduce quantum mechanics more formally. The postulates of quantum mechanics are a list of prescription to summarize: (1) how to describe the *state* of a physical system; (2) how to describe the *measurement* performed on a physical system; (3) how to describe the *evolution* of a physical system.

Postulate 1: States of a Quantum System Each physical system is associated with a complex Hilbert space $\mathcal{H}$ with inner product. The possible states of the physical system correspond to normalized vectors $|\psi\rangle$, i.e., $\langle\psi|\psi\rangle = 1$, which contain all the information about the system. For a composite system we have:

$$|\psi\rangle = |\psi_1\rangle \otimes \ldots \otimes |\psi_N\rangle \in \mathcal{H}, \tag{2.15}$$

where $\mathcal{H} = \mathcal{H}_1 \otimes \ldots \otimes \mathcal{H}_N$ is the tensor product of the Hilbert spaces $\mathcal{H}_k$ associated with the k-th subsystem. If $|\psi\rangle$ and $|\phi\rangle$ are possible states of a quantum system, then any normalized linear superposition $|\Psi\rangle = \alpha|\psi\rangle + \beta|\phi\rangle$, $\langle\Psi|\Psi\rangle = 1$, is an admissible state of the system (note that, in general, $\langle\psi|\phi\rangle \neq 0$, therefore one may have $\langle\Psi|\Psi\rangle = 1$ but $|\alpha|^2 + |\beta|^2 \neq 1$).

Postulate 2: Quantum Measurements Observable quantities are described by Hermitian operators $\hat{A}$, that is $\hat{A} = \hat{A}^\dagger$. The operator $\hat{A}$ admits a spectral decomposition:

$$\hat{A} = \sum_x a_x \hat{P}(a_x) \tag{2.16}$$

in terms of the real eigenvalues a_x, which are the possible values of the observable, where $\hat{P}(a_x) = |u_x\rangle\langle u_x|$ and $\hat{A}|u_x\rangle = a_x|u_x\rangle$. Note that the orthonormal eigenstates $\{|u_x\rangle\}$ form a basis for the Hilbert space. The probability of obtaining the outcome a_x from the measurement of $\hat{A}$ given the state $|\psi\rangle$ is:

$$p(a_x) = \langle\psi|\hat{P}(a_x)|\psi\rangle = |\langle u_x|\psi\rangle|^2, \tag{2.17}$$

and the overall expectation value is:

$$\langle\hat{A}\rangle = \langle\psi|\hat{A}|\psi\rangle. \tag{2.18}$$

This is the *Born rule*, the fundamental recipe to connect the mathematical description of a quantum state $|\psi\rangle$ to the prediction of quantum theory about the results of an experiment. It is now clear that an overall phase does not have a physical meaning: the two states $|\psi\rangle$ and $e^{i\phi}|\psi\rangle$, when inserted in Eqs. (2.17) and (2.18), lead to the same results and, thus, represent the same physical state!

Postulate 3: Dynamics of a Quantum System The dynamical evolution of a physical system from an initial time t_0 to a time $t \geq t_0$ is described by a unitary operator $\hat{U}(t, t_0)$, with $\hat{U}(t, t_0)\hat{U}^\dagger(t, t_0) = \hat{U}^\dagger(t, t_0)\hat{U}(t, t_0) = \hat{\mathbb{1}}$. If $|\psi_{t_0}\rangle$ is the state of the system at time t_0, then at time t we have:

$$|\psi_t\rangle = \hat{U}(t, t_0)|\psi_{t_0}\rangle. \tag{2.19}$$

Furthermore, given $\hat{U}(t, t_0)$ there exists a unique Hermitian operator $\hat{H}$ such that (Stone theorem):

$$\hat{U}(t, t_0) = \exp\left[-i\hat{H}(t - t_0)\right], \tag{2.20}$$

and the form of $\hat{H}$ can be obtained from its identification with the expression for the classical energy of the system, that is the *Hamiltonian* of the system.

2.4 Quantum Two-Level System: Explicit Analysis

Two-level systems are of extreme interest for quantum mechanics and, in particular, for quantum computation. The Hamiltonian of a two-level system can be written as:

$$\hat{H} = \hbar[\omega_0|0\rangle\langle 0| + \omega_1|1\rangle\langle 1| + \gamma|0\rangle\langle 1| + \gamma^*|1\rangle\langle 0|], \tag{2.21}$$

2.4 Quantum Two-Level System: Explicit Analysis

$E_k = \hbar\omega_k$ being the energy of the state $|k\rangle$ associated with the level $k = 0, 1$. Without loss of generality we can assume the coupling constant $\gamma \in \mathbb{R}$. As you will find solving problem 2.3, the two eigenstates of $\hat{H}$ can be written as:

$$|\psi_\pm\rangle = c_{0,\pm}|0\rangle + c_{1,\pm}|1\rangle, \tag{2.22}$$

where

$$c_{0,\pm} = \frac{g}{\sqrt{(E_\pm - E_0)^2 + g^2}}, \tag{2.23a}$$

$$c_{1,\pm} = \frac{E_\pm - E_0}{\sqrt{(E_\pm - E_0)^2 + g^2}}, \tag{2.23b}$$

and

$$E_\pm = \frac{(E_0 + E_1) \pm \sqrt{(\Delta E)^2 + 4g^2}}{2} \tag{2.24}$$

are the corresponding eigenvalues.

Since:

$$\hat{U}(t)|\psi_\pm\rangle = \exp(-i\omega_\pm t)|\psi_\pm\rangle, \tag{2.25}$$

where $\hbar\omega_\pm = E_\pm$, it is straightforward to calculate the time evolution of the computational basis $\{|0\rangle, |1\rangle\}$ (see problem 2.3). Here we explicitly calculate the time evolution of the generic state:

$$|\phi_0\rangle = c_+|\psi_+\rangle + c_-|\psi_-\rangle, \tag{2.26}$$

with $|c_+|^2 + |c_-|^2 = 1$, that reads:

$$|\phi_t\rangle \equiv \hat{U}(t)|\psi_0\rangle, \tag{2.27}$$

$$= e^{-i\omega_+ t}c_+|\psi_+\rangle + e^{-i\omega_- t}c_-|\psi_-\rangle. \tag{2.28}$$

The probability $p(t) = |\langle\phi_0|\phi_t\rangle|^2 = |\langle\phi_0|\hat{U}(t)|\phi_0\rangle|^2$ to find the evolved state in the initial state $|\phi_0\rangle$ at the time t is thus given by:

$$p(t) = 1 - 4|c_+|^2 \underbrace{\left(1 - |c_+|^2\right)}_{|c_-|^2} \sin^2\left(\frac{\Delta\omega t}{2}\right), \tag{2.29}$$

Fig. 2.2 Probability $p(t)$ given in Eq. (2.29) to find an evolved state in the corresponding initial state as a function of $\Delta\omega t$ for $|c_+|^2 = 1/4$ (red, solid line) and $|c_+|^2 = 1/2$ (blue, dashed line). The minimum value of $p(t)$ at $\Delta\omega t = \pi$ is given by $(\Delta E)^2/[4g^2 + (\Delta E)^2]$

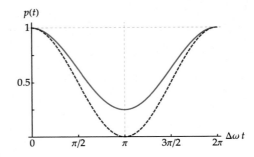

where we introduced $\Delta\omega = \omega_+ - \omega_- = \hbar^{-1}\sqrt{(\Delta E)^2 + 4g^2}$. In Fig. 2.2 we plot $p(t)$ for two different choices of the coefficient c_+ as a function of $\Delta\omega t$.

The last term of Eq. (2.29) represents the interference of the probability amplitudes, whose visibility is:

$$\mathcal{V} = \frac{p_{\max} - p_{\min}}{p_{\max} + p_{\min}}, \qquad (2.30a)$$

$$= \frac{2|c_+|^2 \left(1 - |c_+|^2\right)}{1 - 2|c_+|^2 \left(1 - |c_+|^2\right)}, \qquad (2.30b)$$

where, clearly, $p_{\max} = 1$ and $p_{\min} = 1 - 4|c_+|^2 \left(1 - |c_+|^2\right)$. It is worth noting that the $\mathcal{V}$ reaches its maximum, i.e., 1, if $|c_+|^2 = |c_-|^2 = 1/2$ (see the blue dashed line in Fig. 2.2): the initial state should be a balanced superposition of the eigenstates $|\psi_\pm\rangle$ of the Hamiltonian (2.21), namely:

$$|\phi_0\rangle = \frac{|\psi_+\rangle + e^{i\varphi}|\psi_-\rangle}{\sqrt{2}}. \qquad (2.31)$$

In this case, at times t_n such that $\Delta\omega t_n = 2n\pi$, $n \in \mathbb{N}$, one has $p(t_n) = 0$ and the evolved system is in the state:

$$|\phi_{t_n}\rangle \equiv \left|\phi_0^\perp\right\rangle = \frac{|\psi_+\rangle - e^{i\varphi}|\psi_-\rangle}{\sqrt{2}}, \qquad (2.32)$$

where $\langle\phi_0^\perp|\phi_0\rangle = 0$.

2.5 Structure of 1-Qubit Unitary Transformations

Any 2×2 complex matrix **M** can be written as:

$$\mathbf{M} = r_0 \mathbb{1} + \mathbf{r} \cdot \boldsymbol{\sigma}, \qquad (2.33)$$

2.5 Structure of 1-Qubit Unitary Transformations

where $\boldsymbol{r} = (r_x, r_y, r_z)$, with $r_0, r_k \in \mathbb{C}$, $\boldsymbol{\sigma} = (\sigma_x, \sigma_y, \sigma_x)^T$, σ_k are the Pauli matrices introduced in Eqs. (1.27), $k = x, y, z$, and $\boldsymbol{r} \cdot \boldsymbol{\sigma} = \sum_k r_k \sigma_k$. Here we are interested in unitary transformations, namely, $\mathbf{M}^\dagger \mathbf{M} = \mathbf{M}\mathbf{M}^\dagger = \mathbb{1}$, where $\mathbf{M}^\dagger = r_0^* \mathbb{1} + \boldsymbol{r}^* \cdot \boldsymbol{\sigma}$. Since $\mathbf{M}$ is unitary, also $e^{i\theta}\mathbf{M}$ is unitary, thus we can assume $r_0 \in \mathbb{R}$ without loss of generality.

We have:

$$\mathbf{M}^\dagger \mathbf{M} = (r_0 \mathbb{1} + \boldsymbol{r}^* \cdot \boldsymbol{\sigma})(r_0 \mathbb{1} + \boldsymbol{r} \cdot \boldsymbol{\sigma}) \tag{2.34}$$

that is equivalent to write:

$$\mathbb{1} = r_0^2 \mathbb{1} + r_0 (\boldsymbol{r}^* + \boldsymbol{r}) \cdot \boldsymbol{\sigma} + (\boldsymbol{r}^* \cdot \boldsymbol{\sigma})(\boldsymbol{r} \cdot \boldsymbol{\sigma}). \tag{2.35}$$

By using the identity $(\boldsymbol{a} \cdot \boldsymbol{\sigma})(\boldsymbol{b} \cdot \boldsymbol{\sigma}) = \boldsymbol{a} \cdot \boldsymbol{b} \mathbb{1} + i(\boldsymbol{a} \times \boldsymbol{b}) \cdot \boldsymbol{\sigma}$, $\forall \boldsymbol{a}, \boldsymbol{b} \in \mathbb{C}^3$, we obtain the following two conditions:

$$r_0^2 + \boldsymbol{r}^* \cdot \boldsymbol{r} = 1, \tag{2.36a}$$

$$r_0(\boldsymbol{r}^* + \boldsymbol{r}) + i(\boldsymbol{r}^* \times \boldsymbol{r}) = 0. \tag{2.36b}$$

Since we can write $\boldsymbol{r}^* + \boldsymbol{r} = 2\mathfrak{Re}[\boldsymbol{r}]$ and $i(\boldsymbol{r}^* \times \boldsymbol{r}) = -2\mathfrak{Re}[\boldsymbol{r}] \times \mathfrak{Im}[\boldsymbol{r}]$, Eq. (2.36b) requires $r_0 \mathfrak{Re}[\boldsymbol{r}] = \mathfrak{Re}[\boldsymbol{r}] \times \mathfrak{Im}[\boldsymbol{r}]$, and we have two possibilities. If $r_0 = 0$ and, thus, $\mathfrak{Re}[\boldsymbol{r}]$ is parallel to $\mathfrak{Im}[\boldsymbol{r}]$, then $\boldsymbol{r} = e^{i\phi}\boldsymbol{v}$ with $\boldsymbol{v} \in \mathbb{R}^3$ and, being $\mathbf{M}$ unitary, we can simply write $\boldsymbol{r} = i\boldsymbol{v}$. The second possibility is $r_0 \neq 0$ and, in this case, $\mathfrak{Re}[\boldsymbol{r}]$ should be parallel to $\mathfrak{Re}[\boldsymbol{r}] \times \mathfrak{Im}[\boldsymbol{r}]$. Therefore, $\mathfrak{Re}[\boldsymbol{r}] = 0$ and, again, $\boldsymbol{r} = i\boldsymbol{v}$. Summarizing, for an unitary 2×2 matrix we have:

$$\mathbf{M} = r_0 \mathbb{1} + i\boldsymbol{v} \cdot \boldsymbol{\sigma}, \tag{2.37}$$

where $\boldsymbol{v} \in \mathbb{R}^3$. Furthermore, the condition in Eq. (2.36a) allows us to write:

$$\mathbf{M} = \cos\gamma \, \mathbb{1} + i \sin\gamma \, \boldsymbol{n} \cdot \boldsymbol{\sigma}, \tag{2.38}$$

with $\boldsymbol{n} = \boldsymbol{v}/\sqrt{\boldsymbol{v} \cdot \boldsymbol{v}}$. Finally, we have following useful identity:

$$\exp(i\gamma \, \boldsymbol{n} \cdot \boldsymbol{\sigma}) = \cos\gamma \, \mathbb{1} + i \sin\gamma \, \boldsymbol{n} \cdot \boldsymbol{\sigma}. \tag{2.39}$$

2.5.1 Linear Transformations and Pauli Matrices

The Pauli matrices introduced in Eqs. (1.27) are a basis for 2×2 matrices. Therefore, by using the property $\text{Tr}[\sigma_h \sigma_k] = 2\delta_{hk}$, we can write:

$$\mathbf{M} = \frac{1}{2} \sum_{k=0}^{3} \text{Tr}[\mathbf{M}\sigma_k] \sigma_k, \tag{2.40}$$

where $\sigma_0 = \mathbb{1}$ and $(\sigma_1, \sigma_2, \sigma_3) = (\sigma_x, \sigma_y, \sigma_z)$. The explicit expression of **M** as a function of its matrix elements reads:

$$\mathbf{M} = \begin{pmatrix} m_{00} & m_{01} \\ m_{10} & m_{11} \end{pmatrix}, \tag{2.41a}$$

$$= \frac{m_{00} + m_{11}}{2} \mathbb{1} + \frac{m_{01} + m_{10}}{2} \sigma_x + i \frac{m_{01} - m_{10}}{2} \sigma_y + \frac{m_{00} - m_{11}}{2} \sigma_z. \tag{2.41b}$$

2.6 Quantum States, Density Operator and Density Matrix

Let us consider the following statistical ensemble $\{p_x, |\psi_x\rangle\}$, in which each state $|\psi_k\rangle$ is prepared with probability p_k. Given the observable $\hat{A}$ we have:

$$\langle \hat{A} \rangle = \sum_x p_x \langle \psi_x | \hat{A} | \psi_x \rangle \tag{2.42}$$

and, chosen the orthonormal basis $\{|\phi_s\rangle\}$, we can introduce a resolution of the identity, obtaining:

$$\langle \hat{A} \rangle = \sum_x p_x \langle \psi_x | \hat{A} \left(\sum_s |\phi_s\rangle\langle\phi_s| \right) |\psi_x\rangle,$$

$$= \sum_{x,s} p_x \langle \phi_s | \psi_x \rangle \langle \psi_x | \hat{A} | \phi_s \rangle. \tag{2.43}$$

Rearranging the factors of the last equation, we can write:

$$\langle \hat{A} \rangle = \sum_s \langle \phi_s | \underbrace{\left(\sum_x p_x |\psi_x\rangle\langle\psi_x| \right)}_{\hat{\varrho}} \hat{A} | \phi_s \rangle,$$

$$= \sum_s \langle \phi_s | \hat{\varrho} \hat{A} | \phi_s \rangle \equiv \text{Tr}[\hat{\varrho} \hat{A}]. \tag{2.44}$$

The linear operator $\hat{\varrho}$ is called *density operator*.

More in general, a linear operator:

$$\hat{\varrho} = \sum_{n,m} \varrho_{n,m} |\phi_n\rangle\langle\phi_m|, \tag{2.45}$$

2.6 Quantum States, Density Operator and Density Matrix

with $\varrho_{n,m} = \langle \phi_n | \hat{\varrho} | \phi_m \rangle$, is a density operator describing a physical system if $\hat{\varrho} = \hat{\varrho}^\dagger$, $\hat{\varrho} \geq 0$ and $\mathrm{Tr}[\hat{\varrho}] = 1$.

The matrix ϱ of the coefficients $\varrho_{n,m}$ is the *density matrix* of the physical system. Of course, ϱ is diagonal if we write it in the basis of its eigenstates. For example, the two density operators:

$$\hat{\varrho}_a = \frac{1}{2} \left(|0\rangle\langle 0| + |0\rangle\langle 1| + |1\rangle\langle 0| + |1\rangle\langle 1| \right), \tag{2.46a}$$

$$\hat{\varrho}_b = |+\rangle\langle +|, \tag{2.46b}$$

with $|\pm\rangle = 2^{-1/2}(|0\rangle \pm |1\rangle)$, represent the *same* statistical ensemble written in different basis. In fact the two orthonormal states $|\pm\rangle$ are obtained by applying the Hadamard transformation, which is unitary, to the basis $\{|0\rangle, |1\rangle\}$.

2.6.1 Pure States and Statistical Mixtures

Note that $\hat{\varrho}_a^2 = \hat{\varrho}_a$ while $\hat{\varrho}_c^2 \neq \hat{\varrho}_c$, where $\hat{\varrho}_a$ and $\hat{\varrho}_c$ are given in Eqs. (2.46) and (2.89), respectively. Therefore we also have:

$$\mathrm{Tr}[\hat{\varrho}_a] = \mathrm{Tr}[\hat{\varrho}_a^2] = 1, \tag{2.47}$$

but

$$\mathrm{Tr}[\hat{\varrho}_c^2] = 1/2 < 1. \tag{2.48}$$

Given a density operator $\hat{\varrho}$, in general one has:

$$\mu[\hat{\varrho}] = \mathrm{Tr}[\hat{\varrho}^2] \leq 1, \tag{2.49}$$

where the real, positive quantity $\mu[\hat{\varrho}]$ is the *purity* of the state $\hat{\varrho}$. In the case of a n-dimensional state we find:

$$\frac{1}{n} \leq \mu[\hat{\varrho}] \leq 1. \tag{2.50}$$

If $\mu[\hat{\varrho}] < 1$ then the state is a "statistical mixture", otherwise, i.e., if $\mu[\hat{\varrho}] = 1$, it is "pure". In fact, in the latter case, we can always write $\hat{\varrho} = |\psi\rangle\langle\psi|$. It is now clear that the state $\hat{\varrho}_c$ of Eq. (2.89) is the maximally mixed state for a qubit, i.e., a 2-dimensional state.

2.6.2 Density Operator of a Single Qubit

In the case of a single qubit the density matrix ϱ is a 2×2 matrix and, thus, by means of Eq. (2.40) we can write:

$$\varrho = \frac{1}{2} \left\{ \text{Tr}[\varrho] \mathbb{1} + \text{Tr}[\varrho \, \sigma_x] \, \sigma_x + \text{Tr}[\varrho \, \sigma_y] \, \sigma_y + \text{Tr}[\varrho \, \sigma_z] \, \sigma_z \right\}. \quad (2.51)$$

A similar relation holds for the density operator:

$$\hat{\varrho} = \frac{1}{2} \left\{ \text{Tr}[\hat{\varrho}] \hat{\mathbb{1}} + \text{Tr}[\hat{\varrho} \, \hat{\sigma}_x] \, \hat{\sigma}_x + \text{Tr}[\hat{\varrho} \, \hat{\sigma}_y] \, \hat{\sigma}_y + \text{Tr}[\hat{\varrho} \, \hat{\sigma}_z] \, \hat{\sigma}_z \right\}. \quad (2.52)$$

From now on, we can focus on the matrix representation of the operators, but we have the same result using the operator formalism. Since $\text{Tr}[\hat{\varrho}] = 1$, we find:

$$\varrho = \frac{1}{2} \left(\mathbb{1} + \boldsymbol{r} \cdot \boldsymbol{\sigma} \right), \quad (2.53)$$

where we used the same formalism introduced in Sect. 2.5. Note that, from the physical point of view, the elements of the Bloch vector are the expectations of the Pauli operators, namely, $r_k = \langle \hat{\sigma}_k \rangle = \text{Tr}[\hat{\varrho} \, \hat{\sigma}_k]$, $k = x, y, z$.

Let us now consider ϱ^2, which explicitly reads:

$$\varrho^2 = \frac{1}{4} \left[\mathbb{1} + 2\boldsymbol{r} \cdot \boldsymbol{\sigma} + (\boldsymbol{r} \cdot \boldsymbol{\sigma})(\boldsymbol{r} \cdot \boldsymbol{\sigma}) \right]. \quad (2.54)$$

Since $(\boldsymbol{r} \cdot \boldsymbol{\sigma})(\boldsymbol{r} \cdot \boldsymbol{\sigma}) = \boldsymbol{r} \cdot \boldsymbol{r} \mathbb{1} + i(\boldsymbol{r} \times \boldsymbol{r}) \cdot \boldsymbol{\sigma} = |\boldsymbol{r}|^2 \mathbb{1}$ we have the following expression for the purity:

$$\mu[\hat{\varrho}] = \frac{1}{2} \left(1 + |\boldsymbol{r}|^2 \right), \quad (2.55)$$

and, being $\mu[\varrho] \leq 1$, we have the following condition on the Bloch vector $\boldsymbol{r}$:

$$|\boldsymbol{r}| \leq 1, \quad (2.56)$$

which is needed in order to represent a physical state.

2.7 The Partial Trace

Let $|\psi_{AB}\rangle \in \mathcal{H}_A \otimes \mathcal{H}_B$ and let us consider the measurement of the observable $\hat{A} = \sum_x a_x \hat{P}(a_x)$ on the system A. The overall observable measured on the global

2.7 The Partial Trace

system A–B writes $\hat{A} \otimes \hat{\mathbb{I}}$ and we have the following probability for the outcome a_x (see the Postulate 2 in Sect. 2.3):

$$p(a_x) = \text{Tr}_{AB}\left[\hat{\varrho}_{AB}\,\hat{P}(a_x) \otimes \hat{\mathbb{I}}\right], \tag{2.57}$$

with $\hat{\varrho}_{AB} = |\psi_{AB}\rangle\langle\psi_{AB}|$. As a matter of fact, the Born rule should be valid also for the single system A, thus neglecting system B, namely, we can write:

$$p(a_x) = \text{Tr}_{A}\left[\hat{\varrho}_{A}\,\hat{P}(a_x)\right], \tag{2.58}$$

where $\hat{\varrho}_A$ is the density operator describing the subsystem A. It is possible to show that the *unique map* $\hat{\varrho}_{AB} \to \hat{\varrho}_A$ that allows to maintain the Born rule at the level of the whole system and subsystem is the partial trace:

$$\hat{\varrho}_A = \text{Tr}_B[\hat{\varrho}_{AB}]. \tag{2.59}$$

Note that $\text{Tr}_A[\hat{\varrho}_A] = \text{Tr}_{AB}[\hat{\varrho}_{AB}] = 1$. In fact, by introducing the orthonormal basis $\{|\phi_s^{(K)}\rangle\}$ of the system $K = A, B$, we have:

$$p(a_x) = \text{Tr}_B \text{Tr}_A \left[\hat{\varrho}_{AB}\,\hat{P}(a_x) \otimes \hat{\mathbb{I}}\right]$$
$$= \sum_t \langle\phi_t^{(B)}|\underbrace{\sum_s \langle\phi_s^{(A)}|\hat{\varrho}_{AB}\,\hat{P}(a_x) \otimes \hat{\mathbb{I}}|\phi_s^{(A)}\rangle}_{\text{Tr}_A[\hat{\varrho}_{AB}\,\hat{P}(a_x) \otimes \hat{\mathbb{I}}]}|\phi_t^{(B)}\rangle. \tag{2.60}$$

Due to the linearity, we can exchange the two sums:

$$p(a_x) = \sum_s \langle\phi_s^{(A)}|\underbrace{\sum_t \langle\phi_t^{(B)}|\hat{\varrho}_{AB}\,\hat{P}(a_x) \otimes \hat{\mathbb{I}}|\phi_t^{(B)}\rangle}_{\text{Tr}_B[\hat{\varrho}_{AB}\,\hat{P}(a_x) \otimes \hat{\mathbb{I}}]}|\phi_s^{(A)}\rangle, \tag{2.61}$$

and, rearranging the terms, we find:

$$p(a_x) = \sum_s \langle\phi_s^{(A)}|\underbrace{\sum_t \langle\phi_t^{(B)}|\hat{\varrho}_{AB}\,\hat{\mathbb{I}}|\phi_t^{(B)}\rangle}_{\hat{\varrho}_A \equiv \text{Tr}_B[\hat{\varrho}_{AB}]}\hat{P}(a_x)|\phi_s^{(A)}\rangle$$
$$= \sum_s \langle\phi_s^{(A)}|\hat{\varrho}_A\,\hat{P}(a_x)|\phi_s^{(A)}\rangle \equiv \text{Tr}_A\left[\hat{\varrho}_A\,\hat{P}(a_x)\right]. \tag{2.62}$$

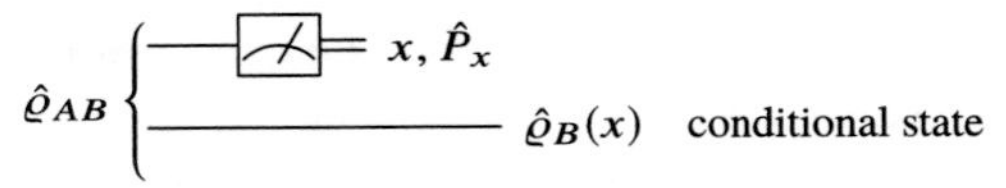

Fig. 2.3 Conditional measurement performed on one qubit of a two-qubit state $\hat{\varrho}_{AB}$. See the text for details

2.7.1 Purification of Mixed Quantum States

Any quantum state $\hat{\varrho}_A$ can be written in the diagonal form choosing its eigenvectors $\left\{\left|\psi_x^{(A)}\right\rangle\right\}$ as the basis for the corresponding Hilbert space $\mathcal{H}_A$, that is

$$\hat{\varrho}_A = \sum_x \lambda_x \left|\psi_x^{(A)}\right\rangle\left\langle\psi_x^{(A)}\right|, \tag{2.63}$$

where $\lambda_x \geq 0$ are the eigenvalues. Let us now consider another Hilbert space $\mathcal{H}_B$ with dimension at least equal to the number of nonzero eigenvalues λ_x and let $\left\{\left|\theta_x^{(B)}\right\rangle\right\}$ a basis of $\mathcal{H}_B$. We have that the following *pure* state:

$$|\Psi_{AB}\rangle = \sum_x \sqrt{\lambda_x} \left|\psi_x^{(A)}\right\rangle\left|\theta_x^{(B)}\right\rangle, \tag{2.64}$$

is such that:

$$\text{Tr}_B\left[|\Psi_{AB}\rangle\langle\Psi_{AB}|\right] = \sum_x \lambda_x \left|\psi_x^{(A)}\right\rangle\left\langle\psi_x^{(A)}\right| = \hat{\varrho}_A, \tag{2.65}$$

that is $|\Psi_{AB}\rangle$ is a *purification* of $\hat{\varrho}_A$.

2.7.2 Conditional States

Figure 2.3 shows a *quantum circuit*[4] in which the qubit belonging to the system A of the input state $\hat{\varrho}_{AB}$ undergoes a projective measurement $\hat{P}_x$. Given the outcome x from the measurement, the conditional state of system B reads:

$$\hat{\varrho}_B(x) = \frac{\text{Tr}_A\left[\hat{P}_x \otimes \hat{\mathbb{I}}\, \hat{\varrho}_{AB}\, \hat{P}_x \otimes \hat{\mathbb{I}}\right]}{p(x)} \tag{2.66}$$

with $p(x) = \text{Tr}\left[\hat{\varrho}_{AB}\, \hat{P}_x \otimes \hat{\mathbb{I}}\right]$.

---

[4] The representation of quantum evolution and measurement by means of quantum circuits will be discussed in the next chapter.

2.8 Entanglement of Two-Qubit States

A pure state of two qubits belonging to the Hilbert space $\mathcal{H}_A \otimes \mathcal{H}_B$ which can be written as the tensor product of the two single-qubit states, namely, $|\psi_A\rangle|\phi_B\rangle$ is called *factorized* or *separable* state. A state which is not separable is called *entangled*, as the following one:

$$|\Psi_{AB}\rangle = \frac{|0_A\rangle|0_B\rangle + |1_A\rangle|1_B\rangle}{\sqrt{2}}, \qquad (2.67)$$

which cannot be written as a tensor product of the two single-qubit states. In particular the state (2.67) is a maximally entangled state.

Entanglement is a key ingredient in many quantum protocols and the characterization of entangled states as well the quantification of this resource is of extreme relevance. A measure $\mathcal{M}_E[\hat{\varrho}_{AB}]$ of the entanglement of the state $\hat{\varrho}_{AB}$ should satisfy the following two conditions:

- $\mathcal{M}_E[\hat{\varrho}_{AB}] = 0 \Leftrightarrow \hat{\varrho}_{AB} = \sum_k p_k \hat{\varrho}_A^{(k)} \otimes \hat{\varrho}_B^{(k)}$, with $p_k \geq 0$ and $\sum_k p_k = 1$ (factorized state);
- given two local unitary operations $\hat{U}_A$ and $\hat{U}_B$ acting the sub-system A and B, respectively, $\mathcal{M}_E[\hat{U}_A \otimes \hat{U}_B \hat{\varrho}_{AB} \hat{U}_A^\dagger \otimes \hat{U}_B^\dagger] = \mathcal{M}_E[\hat{\varrho}_{AB}]$.

2.8.1 Entropy of Entanglement

In the presence of pure states, the simplest measure of entanglement is given by the entropy of entanglement:

$$E(\hat{\varrho}_{AB}) = S[\hat{\varrho}_A] = S[\hat{\varrho}_B], \qquad (2.68)$$

where:

$$S[\hat{\varrho}] = -\text{Tr}[\hat{\varrho} \log_2 \hat{\varrho}] \qquad (2.69)$$

is the von Neumann entropy. In the presence of a pure state $\hat{\varrho} = |\psi\rangle\langle\psi|$, one finds $S[\hat{\varrho}] = 0$. On the other hand, given a N-level system the von Neumann entropy reaches its maximum $S_{\max} = \log_2 N$ for $\varrho = N^{-1}\hat{\mathbb{I}}$, that is the maximally mixed state. Note that, because of the definition of the von Neumann entropy, this measure is independent of the Hilbert space basis and invariant under local unitary operations.

We focus on two two-level systems and start our analysis from the factorized state:

$$|\Psi_{AB}\rangle = \frac{1}{\sqrt{2}}(|0_A\rangle + |1_A\rangle) \otimes \frac{1}{\sqrt{2}}(|0_B\rangle + |1_B\rangle) = \frac{1}{2}\begin{pmatrix}1\\1\\1\\1\end{pmatrix}. \quad (2.70)$$

Since the state (2.70) is a tensor product of two pure states, its entropy of entanglement is null, namely $E(|\Psi_{AB}\rangle) = 0$. Now we consider the two-qubit unitary operation $\text{CPh}(\varphi)$ associated with the following 4×4 matrix (we drop the null elements):

$$\text{CPh}(\varphi) = \begin{pmatrix} 1 & & & \\ & 1 & & \\ & & \cos(\varphi/2) & -\sin(\varphi/2) \\ & & \sin(\varphi/2) & \cos(\varphi/2) \end{pmatrix}, \quad (2.71)$$

$$= \frac{1}{2}(\mathbb{1} + \sigma_z) \otimes \mathbb{1} + \frac{1}{2}(\mathbb{1} - \sigma_z) \otimes \exp\left(-i\frac{\varphi}{2}\sigma_y\right), \quad (2.72)$$

which corresponds to a controlled phase shift: a phase shift φ is applied to the qubit B is the qubit A is the state $|1_A\rangle$. If $\varphi = \pi$, the action of $\text{CPh}(\pi)$ is similar to that of the CNOT, up to a phase [see Eq. (1.19)]. We have:

$$|\Phi_{AB}\rangle \equiv \text{CPh}(\varphi)|\Psi_{AB}\rangle = \frac{1}{2}\begin{pmatrix}1\\1\\c_-\\c_+\end{pmatrix}, \quad (2.73)$$

where $c_\pm = \cos(\varphi/2) \pm \sin(\varphi/2)$. The two sub-systems are described by the density matrices:

$$\varrho_A = \frac{1}{2}\begin{pmatrix}1 & \cos(\varphi/2) \\ \cos(\varphi/2) & 1\end{pmatrix}, \quad \text{and} \quad \varrho_B = \frac{1}{2}\begin{pmatrix}1 - \frac{1}{2}\sin\varphi & \cos^2(\varphi/2) \\ \cos^2(\varphi/2) & 1 + \frac{1}{2}\sin\varphi\end{pmatrix}, \quad (2.74)$$

which both have the following eigenvalues:

$$\lambda_\pm = \frac{1}{2}\left(1 \pm \cos\frac{\varphi}{2}\right). \quad (2.75)$$

2.8 Entanglement of Two-Qubit States

The corresponding entropy of entanglement is:

$$\begin{aligned}E(|\Phi_{AB}\rangle) = &-\frac{1}{2}\left(1-\cos\frac{\varphi}{2}\right)\log_2\left[\frac{1}{2}\left(1-\cos\frac{\varphi}{2}\right)\right]\\&-\frac{1}{2}\left(1+\cos\frac{\varphi}{2}\right)\log_2\left[\frac{1}{2}\left(1+\cos\frac{\varphi}{2}\right)\right],\end{aligned} \quad (2.76)$$

which vanishes for $\varphi = 0, 2\pi$ and reaches the maximum $E(|\Phi_{AB}\rangle) = \log_2 2 = 1$ for $\phi = \pi$. It is then clear that for $\varphi \neq 0, 2\pi$ the operation CPh(φ) is an *entangling gate*.

2.8.2 Concurrence

Another measure of entanglement is the concurrence. Given the two-qubit pure state:

$$|\psi_{AB}\rangle = \sum_{x,y}\alpha_{xy}|x_A\rangle|y_B\rangle, \quad (2.77)$$

with $\alpha_{xy} \in \mathbb{C}$, $x, y \in \{0, 1\}$, and $\sum_{x,y}|\alpha_{xy}|^2 = 1$, the concurrence is defined as:

$$C(|\psi_{AB}\rangle) = 2|\alpha_{00}\alpha_{11} - \alpha_{01}\alpha_{10}|. \quad (2.78)$$

If $C = 0$, the state is factorized, whereas if $C > 0$, the state is entangled. Since:

$$\begin{aligned}4|\alpha_{00}\alpha_{11} - \alpha_{01}\alpha_{10}|^2 &= 4\left[|\alpha_{00}\alpha_{11}|^2 + |\alpha_{01}\alpha_{10}|^2 - \alpha_{00}\alpha_{11}\alpha_{01}^*\alpha_{10}^* - \alpha_{00}^*\alpha_{11}^*\alpha_{01}\alpha_{10}\right]\\&= 4\left\{\left(|\alpha_{00}|^2 + |\alpha_{01}|^2\right)\left(|\alpha_{10}|^2 + |\alpha_{11}|^2\right) - |\alpha_{00}\alpha_{01}^* + \alpha_{01}\alpha_{11}^*|^2\right\}\\&\leq 4\left(|\alpha_{00}|^2 + |\alpha_{01}|^2\right)\left[1 - \left(|\alpha_{00}|^2 + |\alpha_{01}|^2\right)\right] \leq 1,\end{aligned} \quad (2.79)$$

we have $0 \leq C(|\psi_{AB}\rangle) \leq 1$.

The concurrence (2.78) can be written as a function of the purity of the subsystem states. For instance, the density matrix of the sub-system A of the state in Eq. (2.77) reads:

$$\varrho_A = \begin{pmatrix} |\alpha_{00}|^2 + |\alpha_{01}|^2 & \alpha_{00}\alpha_{01}^* + \alpha_{01}\alpha_{11}^* \\ \alpha_{00}^*\alpha_{01} + \alpha_{01}^*\alpha_{11} & |\alpha_{10}|^2 + |\alpha_{11}|^2 \end{pmatrix}, \quad (2.80)$$

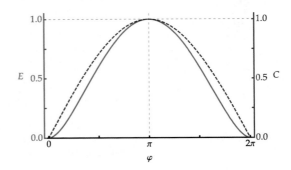

Fig. 2.4 Plots of the entropy of entanglement E (red solid line, left vertical axis) and concurrence C (blue dashed line, right vertical axis) of the state $|\Phi_{AB}\rangle$ of Eq. (2.73)

therefore we have:

$$C(|\psi_{AB}\rangle) = 2\sqrt{\det[\varrho_A]}. \tag{2.81}$$

Furthermore, using the results of Sect. 2.6.2, we can write $\varrho_A = \frac{1}{2}(\mathbb{1} + \boldsymbol{r}_A \cdot \boldsymbol{\sigma})$, where $|\boldsymbol{r}_A|^2 = 2\text{Tr}[\varrho_A^2] - 1$, and, thus, we obtain the following expression for the concurrence:

$$C(|\psi_{AB}\rangle) = \sqrt{1 - |\boldsymbol{r}_A|^2}. \tag{2.82}$$

In Fig. 2.4 we plot the entropy of entanglement and the concurrence of the state (2.73). It is clear that the numerical values of the two entanglement measures are different, but they reach the maximum ($E = C = 1$) in the presence of a maximally entangled state while they both vanish for a factorized state.

Though the entropy of entanglement is a good measure only in the presence of pure two-qubit states, the concurrence can be extended also to mixed states. In this case, given the two-qubit density operator $\hat{\varrho}_{AB}$, the concurrence is given by:

$$C(\hat{\varrho}_{AB}) = \max(0, \lambda_1 - \lambda_2 - \lambda_3 - \lambda_4), \tag{2.83}$$

where $\lambda_1 \geq \lambda_2 \geq \lambda_3 \geq \lambda_4$ are the eigenvalues of the operator:

$$\hat{R} = \sqrt{\sqrt{\hat{\varrho}_{AB}}\,\hat{\varrho}'_{AB}\sqrt{\hat{\varrho}_{AB}}}, \tag{2.84}$$

with $\hat{\varrho}'_{AB} = \hat{\sigma}_y \otimes \hat{\sigma}_y \hat{\varrho}^*_{AB} \hat{\sigma}_y \otimes \hat{\sigma}_y$.

2.9 Quantum Measurements and POVMs

In the previous sections we have seen that a *projective* measurement with outcome x is described by the operators $\hat{P}_x = \hat{P}_x^2 \geq 0$, that is $\hat{P}_x$ is a positive operator. Given the state $\hat{\varrho}$, we have the following expressions for the probability of the outcome x

and the corresponding conditional state $\hat{\varrho}_x$:

$$p(x) = \text{Tr}\left[\hat{P}_x\, \hat{\varrho}\, \hat{P}_x\right] = \text{Tr}\left[\hat{\varrho}\, \hat{P}_x^2\right] = \text{Tr}\left[\hat{\varrho}\, \hat{P}_x\right], \qquad (2.85)$$

and:

$$\hat{\varrho}_x = \frac{\hat{P}_x\, \hat{\varrho}\, \hat{P}_x}{p(x)}, \qquad (2.86)$$

respectively.

A generalized measurement, not described by projectors, is a positive operator-valued measure (POVM), i.e., a set of positive operators $\{\hat{\Pi}_x\}$, $\hat{\Pi}_x \geq 0$, such that $\sum_x \hat{\Pi}_x = \hat{\mathbb{I}}$. In this case we can have $\hat{\Pi}_x^2 \neq \hat{\Pi}_x$ and the probability of the outcome x and the corresponding conditional state $\hat{\varrho}_x$ read:

$$p(x) = \text{Tr}\left[\hat{\varrho}\, \hat{\Pi}_x\right] = \text{Tr}\left[\hat{M}_x\, \hat{\varrho}\, \hat{M}_x^\dagger\right], \qquad (2.87)$$

where $\hat{\Pi}_x = \hat{M}_x^\dagger \hat{M}_x$ or $\hat{M}_x = \sqrt{\hat{\Pi}_x}$, and:

$$\hat{\varrho}_x = \frac{\hat{M}_x\, \hat{\varrho}\, \hat{M}_x^\dagger}{p(x)}, \qquad (2.88)$$

respectively.

Problems

2.1. Exploiting the completeness relation $\sum_x |x\rangle\langle x| = \hat{\mathbb{I}}$, write the expansion of $|\psi\rangle$ in the basis $\{|x\rangle\}$.

2.2. Exploiting the completeness relation $\sum_x |x\rangle\langle x| = \hat{\mathbb{I}}$, write the expansion of a linear operator $\hat{A}$ in the basis $\{|x\rangle\}$.

2.3. ♣ (Two-level system) Given the (quantum) Hamiltonian:

$$\hat{H} = \hbar[\omega_0|0\rangle\langle 0| + \omega_1|1\rangle\langle 1| + \gamma(|1\rangle\langle 0| + |0\rangle\langle 1|)],$$

where we used the computational basis $\{|0\rangle, |1\rangle\}$, find the eigenvalues and the eigenstates of $\hat{H}$ and calculate:

$$\hat{U}(t)|1\rangle = \exp\left(-i\hat{H}t/\hbar\right)|1\rangle.$$

(Hint: express the Hamiltonian in its matrix form...)

2.4. ♣ Prove Eq. (2.39) by using the expansion:

$$\exp(i\gamma\, \boldsymbol{n}\cdot\boldsymbol{\sigma}) = \sum_{k=0}^{\infty} \frac{(i\gamma)^k}{k!} (\boldsymbol{n}\cdot\boldsymbol{\sigma})^k.$$

2.5. Write the density matrices of the states in Eqs. (2.46) in the computational basis $\{|0\rangle, |1\rangle\}$ and in the transformed basis $|\pm\rangle$.

2.6. Write the density operator and the density matrix of the state

$$\hat{\varrho}_c = \frac{1}{2}\left(|+\rangle\langle+| + |-\rangle\langle-|\right), \qquad (2.89)$$

in the computational basis $\{|0\rangle, |1\rangle\}$.

2.7. Given the density operator $\hat{\varrho}_{AB}$ describing the state of a bipartite system A–B and the observable $\hat{A} = \sum_x a_x\, \hat{P}(a_x)$ on the system A, show that $\langle \hat{A} \rangle = \mathrm{Tr}_A\left[\hat{\varrho}_A\, \hat{A}\right]$, where $\hat{\varrho}_A = \mathrm{Tr}_B[\hat{\varrho}_{AB}]$.

2.8. Given the following 3-qubit state (the bit order 1-2-3 is from left to right as usual):

$$|\psi\rangle = \alpha|010\rangle - \beta|101\rangle + \gamma|110\rangle, \qquad (2.90)$$

with $|\alpha|^2 + |\beta|^2 + |\gamma|^2 = 1$, write the conditional state of qubits 2 and 3 and the corresponding probability of obtaining it, when one performs a measurement involving only the qubit 1. (Note that the final state should be normalized!)

Further Readings

M.A. Nielsen, I.L. Chuang, *Quantum Computation and Quantum Information* (Cambridge University Press, 2010) – Chapter 2

M.G.A. Paris, The modern tools of quantum mechanics. Eur. Phys. J. Special Topics **203**, 61–86 (2012)

S. Stenholm, K.A. Suominen, *Quantum Approach to Informatics* (Wiley-Interscience, 2005) – Chapter 2

Quantum Mechanics as Computation 3

Abstract

In this chapter we introduce the basic framework of quantum computation as an abstract extension of the classical logic. Quantum logic gates and their quantum circuit representations are given. Applications to Bell state generation and Bell state measurement as well as to the quantum teleportation are provided. Furthermore, we address the Deutsch, the Deutsch–Jozsa and the Bernstein–Vazirani algorithms. We close the chapter with some considerations about the universality of single-qubit gates and CNOT and we state the Gottesman–Knill theorem concerning the efficient simulation of quantum computation on classical computers.

3.1 Quantum Logic Gates

A quantum logic gate transforms an input qubit state as that given in Eq. (2.7) into an output state $|\psi'\rangle = \alpha'|0\rangle + \beta'|1\rangle$. Since the condition $|\alpha'|^2 + |\beta'|^2 = 1$ should be still satisfied, it is possible to show that the action of any quantum logic gate can be represented by a *linear unitary transformation* associated with a unitary operator $\hat{U}$, namely:

$$|\psi\rangle \to |\psi'\rangle \equiv \hat{U}|\psi\rangle, \tag{3.1}$$

where $\hat{U}^\dagger \hat{U} = \hat{U}\hat{U}^\dagger = \hat{\mathbb{I}}$. Being $\hat{U}$ unitary, not only the normalization of the qubit state is preserved during the transformation, but the operation is intrinsically *reversible*. In Fig. 3.1 the unitary transformation (3.1) is schematically represented by means of a *quantum circuit*: the horizontal lines are "wires" representing the time evolution (from left to right), and they connect the "gates", represented by means of boxes labeled by the corresponding unitary operator.

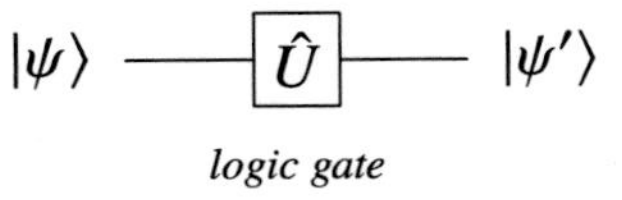

$|\psi\rangle$ —[$\hat{U}$]— $|\psi'\rangle$

logic gate

Fig. 3.1 Example of a simple quantum circuit involving a single input qubit $|\psi\rangle$ and a unitary (quantum) logic gate $\hat{U}$: $|\psi'\rangle$ corresponds to the output state

Fig. 3.2 Quantum circuit for the NOT acting on: (**a**) the bit $|x\rangle$; (**b**) the qubit $|\psi\rangle = \alpha|0\rangle + \beta|1\rangle$

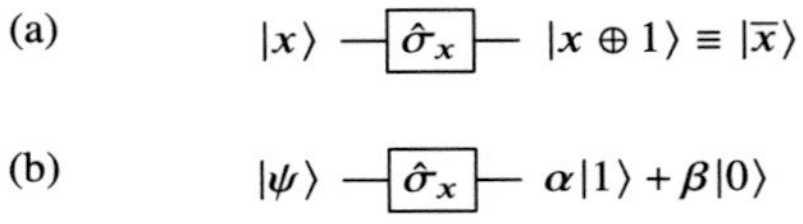

(a) $|x\rangle$ —[$\hat{\sigma}_x$]— $|x \oplus 1\rangle \equiv |\bar{x}\rangle$

(b) $|\psi\rangle$ —[$\hat{\sigma}_x$]— $\alpha|1\rangle + \beta|0\rangle$

Fig. 3.3 Quantum circuit for the Hadamard transformation: (**a**) action of **H** on a single bit $|x\rangle$; (**b**) action of **H** on the qubit $\alpha|0\rangle + \beta|1\rangle$

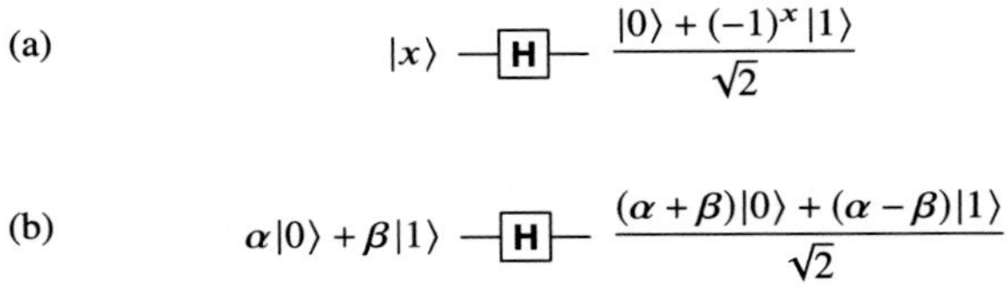

(a) $|x\rangle$ —[H]— $\dfrac{|0\rangle + (-1)^x |1\rangle}{\sqrt{2}}$

(b) $\alpha|0\rangle + \beta|1\rangle$ —[H]— $\dfrac{(\alpha+\beta)|0\rangle + (\alpha-\beta)|1\rangle}{\sqrt{2}}$

3.1.1 Single Qubit Gates

In Chap. 1 we explained that the only reversible classical operation is the NOT gate. In the quantum logic scenario it is represented by the Pauli matrix $\hat{\sigma}_x$ and the corresponding quantum circuit is sketched in Fig. 3.2. Note that due to the linearity of the transformation we have:

$$\hat{\sigma}_x(\alpha|0\rangle + \beta|1\rangle) = \alpha\hat{\sigma}_x|0\rangle + \beta\hat{\sigma}_x|1\rangle \tag{3.2}$$

$$= \alpha|1\rangle + \beta|0\rangle, \tag{3.3}$$

as illustrated in Fig. 3.2b.

In general, a single qubit gate is a linear combination of the Pauli operators. Since any unitary transformation acting on a qubit can be seen as a quantum logic gate, we have infinite single-qubit gates!

Hadamard Transformation In particular, the gate associated with the Hadamard transformation $\mathbf{H} = \frac{1}{\sqrt{2}}(\hat{\sigma}_x + \hat{\sigma}_z)$ defined in Eq. (1.31) not only makes sense (now superpositions of qubit states are allowed!), but it transforms a bit $|x\rangle$ into a superposition and, as we will see, this is a key ingredient of many quantum algorithms. In Fig. 3.3 we can see the schematic representation of the action of **H** on a bit and on a qubit, respectively.

3.1 Quantum Logic Gates

Phase Shift Gate The Pauli operator $\hat{\sigma}_z$ adds a π phase shift between the computational states $|0\rangle$ and $|1\rangle$, since $\hat{\sigma}_z|x\rangle = e^{i\pi x}|x\rangle$. More in general, the phase shift gate acts as the phase shift operator:

$$e^{-i\phi\hat{\sigma}_z} = \cos\phi\,\hat{\mathbb{1}} - i\sin\phi\,\hat{\sigma}_z \to \begin{pmatrix} e^{-i\phi} & 0 \\ 0 & e^{i\phi} \end{pmatrix} = e^{-i\phi}\begin{pmatrix} 1 & 0 \\ 0 & e^{i2\phi} \end{pmatrix}, \quad (3.4)$$

which adds a relative phase shift 2ϕ between the computational basis states.

$\hat{T}$ Gate or $\frac{\pi}{8}$ Gate This gate, usually referred to as $\hat{T}$ gate, represents the action of a phase shift gate with $\phi = \pi/8$, namely:

$$\hat{T} \to T = \begin{pmatrix} 1 & 0 \\ 0 & e^{i\pi/4} \end{pmatrix} = e^{i\pi/8}\begin{pmatrix} e^{-i\pi/8} & 0 \\ 0 & e^{i\pi/8} \end{pmatrix}. \quad (3.5)$$

Phase Gate There are two important gates that can be built starting from the T gate, namely:

$$S = T^2 = \begin{pmatrix} 1 & 0 \\ 0 & i \end{pmatrix}, \quad \text{(phase gate)} \quad (3.6)$$

and:

$$T^4 = \begin{pmatrix} 1 & 0 \\ 0 & -1 \end{pmatrix} \to \hat{\sigma}_z. \quad (3.7)$$

The phase gate S, as a stand alone gate, is justified in order to implement fault-tolerant universal quantum computation (see Sect. 8.4).

3.1.2 Single Qubit Gates and Bloch Sphere Rotations

As a single-qubit pure state can be represented as a point on the Bloch sphere (see Sect. 2.2.1), the action of a quantum gate maps point to point and, thus, can be written as the unitary transformation $U = e^{i\alpha}\mathcal{R}_n(\theta)$, where:

$$\mathcal{R}_n(\theta) = \exp(i\theta \boldsymbol{n} \cdot \boldsymbol{\sigma}), \quad (3.8)$$

is a rotation of 2θ around the unit vector $\boldsymbol{n} \in \mathbb{R}^3$. Due to the properties of the rotations, we can decompose $\mathcal{R}_n(\theta)$ as the combination of rotations around the principal axes z and y axis (or, analogously, x and y). Therefore, the unitary transformation U can be written as:

$$U = e^{i\alpha} \mathcal{R}_z(\beta) \mathcal{R}_y(\gamma) \mathcal{R}_z(\delta), \qquad (3.9)$$

where the values of the angles β, γ and δ depend on $\boldsymbol{n}$ and θ.

3.1.3 Two-Qubit Gates: The CNOT Gate

In Chap. 1 we have seen that any logical or arithmetical function can be computed from the composition of NOR or NAND two-bit gates, which are thus universal gates. However, these operators are not reversible and, thus, they cannot be represented by means of unitary operators. The irreversibility, in fact, can be seen as a loss of information.

The prototypical multiple qubit gate is the CNOT gate we introduced in Sect. 1.4.2 and whose quantum circuit is shown in Fig. 3.4 for what concerns the action of $\mathbf{C}_{10}$ and in Fig. 3.5 for $\mathbf{C}_{01}$. In Fig. 3.6 we show the quantum circuit of a CNOT gate, which changes the value of the target qubit if the control is in the state $|0\rangle$: in this case the full circle on the control wire is substituted by an open one. When we will refer to this gate we will use the symbol $\overline{\text{CNOT}}$ to indicate that the control should be $|0\rangle$ to change the state of the target. Of course, the action of a

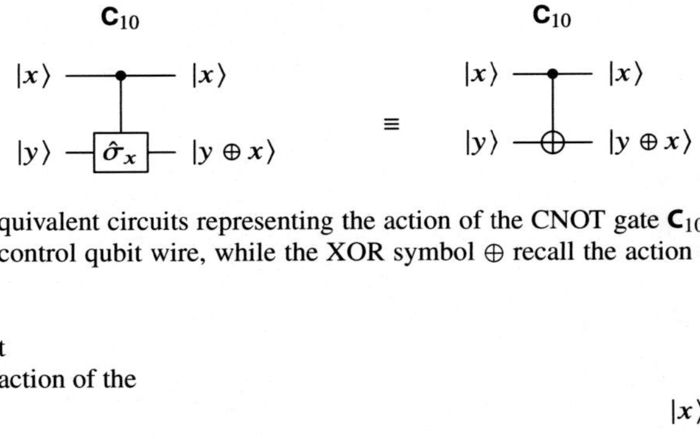

Fig. 3.4 Two equivalent circuits representing the action of the CNOT gate $\mathbf{C}_{10}$. The filled circle is placed on the control qubit wire, while the XOR symbol $\oplus$ recall the action of the gate on the target qubit

Fig. 3.5 Circuit representing the action of the CNOT gate $\mathbf{C}_{01}$

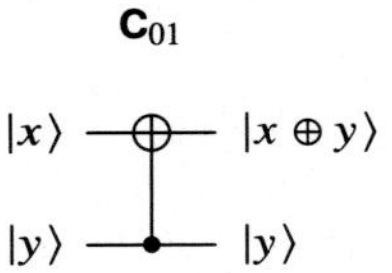

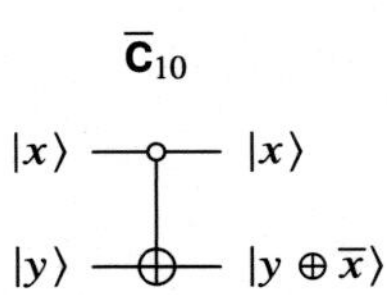

Fig. 3.6 Circuit representing the action of a $\overline{\text{CNOT}}$ gate $\overline{\mathbf{C}}_{10}$, namely a CNOT which changes the value of the target if the control is in the state $|0\rangle$

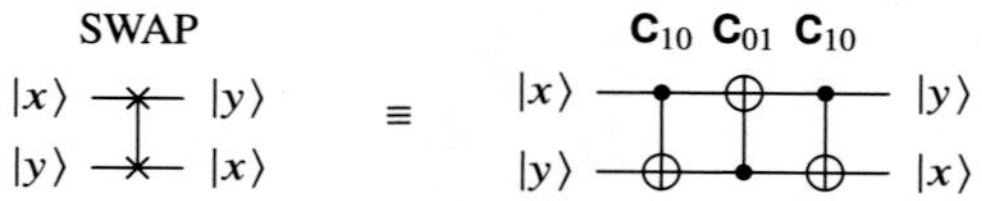

3.1 Quantum Logic Gates

Fig. 3.7 Quantum circuit representing the SWAP operation acting on two qubits composed from three CNOT gates

$\overline{\text{CNOT}}$ can be represented as:

$$\hat{\sigma}_x \otimes \hat{\mathbb{1}} \, \mathbf{C}_{10} \, \hat{\sigma}_x \otimes \hat{\mathbb{1}} \equiv \overline{\mathbf{C}}_{10}, \qquad (3.10)$$

or

$$\hat{\mathbb{1}} \otimes \hat{\sigma}_x \, \mathbf{C}_{01} \, \hat{\mathbb{1}} \otimes \hat{\sigma}_x \equiv \overline{\mathbf{C}}_{01}. \qquad (3.11)$$

It is worth noting that CNOT is a reversible operation on two qubits. In the next sections we will see how any multiple qubit gate may be composed from CNOT and single-qubit gates thus leading to the universal quantum computation. Figure 3.7 shows the quantum circuit of the SWAP operation and its equivalent realization based on three CNOT gates [see also Eq. (1.33)].

The unitary matrix associated with the CNOT (from now on we consider the first qubit as control) reads:

$$U_{\text{CNOT}} = \begin{pmatrix} \mathbb{1} & 0 \\ 0 & \sigma_x \end{pmatrix}. \qquad (3.12)$$

More in general, the unitary matrix cU describing the conditional application of a unitary transformation U to a qubit, namely, $cU|x\rangle|y\rangle = \mathbb{1} \otimes U^x|x\rangle|y\rangle$ writes:

$$cU = \begin{pmatrix} \mathbb{1} & 0 \\ 0 & U \end{pmatrix}. \qquad (3.13)$$

How can we implement the two-qubit gate cU with single-qubit gates and CNOT? We assume that U can be recast in the form (3.9) and introduce the three auxiliary unitary gates:

$$U_A = \mathcal{R}_z(\beta) \mathcal{R}_y\left(\frac{\gamma}{2}\right), \qquad (3.14\text{a})$$

$$U_B = \mathcal{R}_y\left(-\frac{\gamma}{2}\right) \mathcal{R}_z\left(-\frac{\delta + \beta}{2}\right), \qquad (3.14\text{b})$$

$$U_C = \mathcal{R}_z\left(\frac{\delta - \beta}{2}\right). \qquad (3.14\text{c})$$

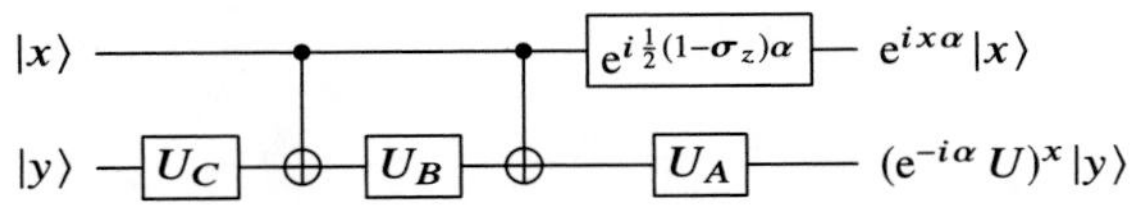

Fig. 3.8 Quantum circuit acting as a cU, where $U_A\sigma_x U_B \sigma_x U_C = e^{-i\alpha}U$

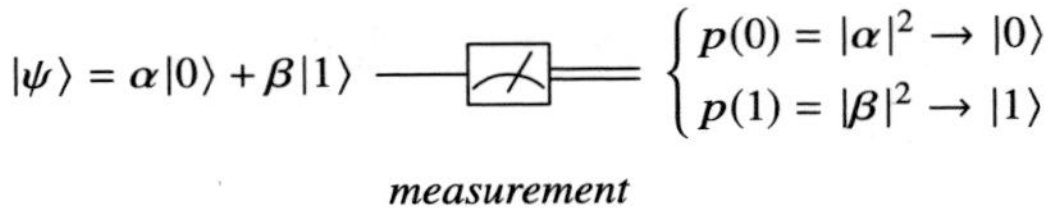

Fig. 3.9 Circuit representing the measurement on the qubit $|\psi\rangle$: though the input is a superposition state, the outcome is either $|0\rangle$, with probability $p(0) = |\alpha|^2$, or $|1\rangle$, with probability $p(1) = |\beta|^2$

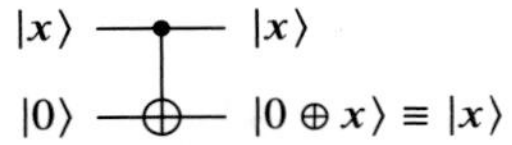

Fig. 3.10 CNOT gate acting as a cloner of the classical bit $|x\rangle$

such that $U_A U_B U_C = \mathbb{1}$. Furthermore, since $\sigma_x \sigma_z \sigma_x = -\sigma_z$ and $\sigma_x \sigma_y \sigma_x = -\sigma_y$, we also have:

$$\sigma_x U_B \sigma_x = \mathcal{R}_y\left(\frac{\gamma}{2}\right) \mathcal{R}_z\left(\frac{\delta+\beta}{2}\right), \tag{3.15}$$

and, thus, $U_A \sigma_x U_B \sigma_x U_C = e^{-i\alpha} U$. The quantum circuit implementing the cU is reported in Fig. 3.8.

3.2 Measurement on Qubits

As we mentioned, the measurement is a critical point. As sketched in Fig. 3.9, the result of a measurement on the qubit (2.7) is a single bit $|0\rangle$ or $|1\rangle$ (the double line after the "meter" represents the classical wire carrying one bit of classical information) with a probability given by $|\alpha|^2$ and $|\beta|^2$, respectively. As a matter of fact, during the measurement process performed onto a qubit there is a (huge!) loss of information, which makes the measurement an irreversible process.

3.3 Applications and Examples

3.3.1 CNOT and No-Cloning Theorem

One of the peculiar aspects of quantum information is that an unknown quantum state cannot be perfectly cloned. This is a consequence of the linear nature of the operators acting on the quantum states.

In Fig. 3.10 it is shown how a CNOT gate can be used to clone a (classical) bit $|x\rangle$, $x = 0, 1$. In this case the state of the input bit $|0\rangle$ is converted into the state $|x\rangle$,

3.3 Applications and Examples

so that the whole process can be summarized as $|x\rangle|0\rangle \to |x\rangle|x\rangle$: we end up with two copies of $|x\rangle$. However, if we try to use the same circuit to clone the qubit $|\psi\rangle$ of Eq. (2.7), we obtain:

$$\mathbf{C}_{10}|\psi\rangle|0\rangle = \alpha|0\rangle|0\rangle + \beta|1\rangle|1\rangle,$$

which is an entangled state (if $\alpha, \beta \neq 0$) and it is indeed different form the desired state:

$$|\psi\rangle|\psi\rangle = \alpha^2|0\rangle|0\rangle + \alpha\beta(|0\rangle|1\rangle + |1\rangle|0\rangle) + \beta^2|1\rangle|1\rangle, \tag{3.16}$$

unless α or β vanishes, but this is exactly the classical case depicted in Fig. 3.10!

3.3.2 Bell States and Bell Measurement

As we have seen in Sect. 2.8 the pure state:

$$|\beta_{00}\rangle = \frac{|0\rangle|0\rangle + |1\rangle|1\rangle}{\sqrt{2}}, \tag{3.17}$$

is entangled since it cannot be written as a tensor product of the two single-qubit states. The state (3.17) is one of the four maximally entangled "Bell states":

$$|\beta_{xy}\rangle = \frac{|0\rangle|y\rangle + (-1)^x|1\rangle|\bar{y}\rangle}{\sqrt{2}}, \tag{3.18a}$$

$$= \frac{1}{\sqrt{2}} \sum_{M=0,1} (\hat{\sigma}_z)^{Mx} \otimes \hat{\mathbb{I}} |M\rangle|y \otimes M\rangle \tag{3.18b}$$

which can be produced starting from the separable state $|x\,y\rangle$ as depicted in Fig. 3.11. Note that the Bell states are a basis for the two-qubit space.

The circuit to generate the Bell states is reversible and its inverse can be used to transform the Bell basis into the usual two-qubit computational basis $\{|0\rangle|0\rangle, |0\rangle|1\rangle, |1\rangle|0\rangle, |1\rangle|1\rangle\}$, as sketched in Fig. 3.12. We can also expand the elements of the computational basis as a superposition of the Bell states, namely:

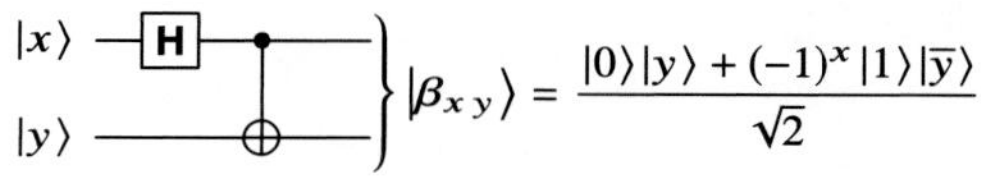

Fig. 3.11 Quantum circuit to generate the Bell state $|\beta_{xy}\rangle$ from the separable state $|x\rangle|y\rangle$

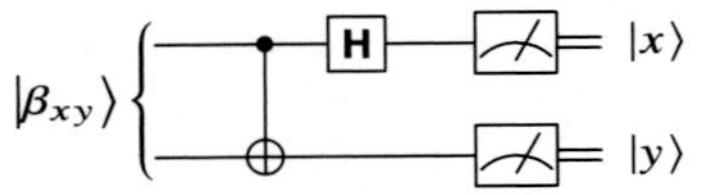

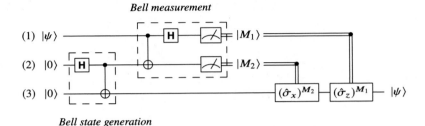

Fig. 3.12 Quantum circuit to perform the Bell measurement: the maximally entangled state $|\beta_{xy}\rangle$ is transformed into the separable state $|x\,y\rangle$ and then measured

Fig. 3.13 Quantum circuit to perform quantum teleportation

$$|x\rangle|y\rangle = \frac{1}{\sqrt{2}} \sum_{M=0,1} (-1)^{Mx} \otimes \hat{\mathbb{I}} \,|\beta_{M\,x\oplus y}\rangle. \qquad (3.19)$$

We use both the Bell generation and the Bell measurement in the next section to implement the so-called "quantum teleportation" protocol.

3.3.3 Quantum Teleportation

As we pointed out, if we measure in the computational basis $\{|0\rangle, |1\rangle\}$ a qubit in a unknown quantum state, we will loose in general any information about it, obtaining as outcome just a classical bit $|x\rangle$ with a certain probability. However, it is sometimes necessary to *transfer* the state of a qubit from one part of a quantum computer to another. In this case, the state can be *teleported*, i.e., the unknown state $|\psi\rangle = \alpha|0\rangle + \beta|1\rangle$ of an input qubit can be reconstructed on a target qubit. The teleportation protocol requires two bits of classical information and a maximally entangled state.

In Fig. 3.13 we sketched the quantum circuit to implement quantum teleportation. The protocol takes as input the three-qubit state $|\psi\rangle|0\rangle|0\rangle$. The first step is to create an entangled state: following the procedure described in Sect. 3.3.2, we create the Bell state $|\beta_{00}\rangle$ on the qubits 2 and 3 (see Fig. 3.13): this furnish the entanglement resource. At this stage the overall three-qubit state reads:

$$|\psi\rangle|\beta_{00}\rangle = \frac{1}{\sqrt{2}} \left[\alpha|0\rangle|0\rangle|0\rangle + \beta|1\rangle|1\rangle|1\rangle + \alpha|0\rangle|1\rangle|1\rangle + \beta|1\rangle|0\rangle|0\rangle \right], \qquad (3.20a)$$

3.3 Applications and Examples

$$= \frac{1}{\sqrt{2}}\Big[|0\rangle|0\rangle(\alpha|0\rangle) + |1\rangle|1\rangle(\beta|1\rangle)$$

$$+ |0\rangle|1\rangle(\alpha\,\hat{\sigma}_x|0\rangle) + |1\rangle|0\rangle(\beta\,\hat{\sigma}_x|1\rangle)\Big]. \quad (3.20b)$$

Now we should perform the Bell measurement (see again Sect. 3.3.2) on qubits 1 and 2 by applying the gate $\mathbf{C}_{12}$ followed by $\mathbf{H}$ acting on qubit 1 and then measuring the two qubits in the computational basis. Since:

$$\left(\mathbf{H}\otimes\hat{\mathbb{I}}\right)\mathbf{C}_{12}|x\rangle|y\rangle = \frac{1}{\sqrt{2}}\sum_{M_1=0,1}(-1)^{M_1 x}|M_1\rangle|y\oplus x\rangle \quad (3.21a)$$

$$= \frac{1}{\sqrt{2}}\sum_{M_1=0,1}\Big[(\hat{\sigma}_z)^{M_1 x}|M_1\rangle\Big]|y\oplus x\rangle, \quad (3.21b)$$

after these transformations (but before the measurement!) the three-qubit state can be written in the following compact form:

$$\left(\mathbf{H}\otimes\hat{\mathbb{I}}\right)\mathbf{C}_{12}|\psi\rangle|\beta_{00}\rangle = \frac{1}{2}\sum_{M_1=0,1}\sum_{M_2=0,1}|M_1\rangle|M_2\rangle\Big[(\hat{\sigma}_x)^{M_2}(\hat{\sigma}_z)^{M_1}(\alpha|0\rangle + \beta|1\rangle)\Big], \quad (3.22)$$

where we used the identity (note the at the l.h.s. the operator $\hat{\sigma}_z$ act on the first qubit, whereas at the r.h.s. it acts on the third qubit):

$$\sum_{M_1=0,1}\Big[(\hat{\sigma}_z)^{M_1}|M_1\rangle\Big]|M_2\rangle\Big[(\hat{\sigma}_x)^{M_2}|1\rangle\Big] = \sum_{M_1=0,1}|M_1\rangle|M_2\rangle\Big[(\hat{\sigma}_x)^{M_2}(\hat{\sigma}_z)^{M_1}|1\rangle\Big], \quad (3.23)$$

with $M_2 = 0, 1$. It is now clear the a measurement carried out on qubits 1 and 2 with outcomes $|M_1\rangle$ and $|M_2\rangle$, respectively, leave the qubit 3 in the state:

$$(\hat{\sigma}_x)^{M_2}(\hat{\sigma}_z)^{M_1}(\alpha|0\rangle + \beta|1\rangle) \equiv (\hat{\sigma}_x)^{M_2}(\hat{\sigma}_z)^{M_1}|\psi\rangle. \quad (3.24)$$

Thus, in order to reconstruct the state of the input qubit onto the qubit 3 we should apply to Eq. (3.24) the unitary transformation $(\hat{\sigma}_x)^{M_2}(\hat{\sigma}_z)^{M_1}$.

It is worth noting that:

- only information is teleported, not matter;
- the input state is lost during the measurement (no-cloning theorem holds);
- no information about the input state is acquired through the measurement (the four outcomes $|M_1 M_2\rangle$ do not contain any information about α an d β since they occur with the same probability, i.e., 25%);

- the teleportation protocol is not instantaneous (one should send to the receiver by a classical channels the information about the two output classical bits $|M_1\rangle$ and $|M_2\rangle$);
- in order to reconstruct the state of a qubit we need two bits of classical information and the entanglement resource.

3.4 The Standard Computational Process

The goal of a computational process is to calculate the values $f(x)$ of some specified function f where x is encoded in the computational-basis state of n qubits, with an accuracy which increases with n.

Since a quantum computer works with reversible operations, while $f(x)$ in general is not, we should specify x and $f(x)$ as an n-bit and m-bit integers, respectively. Then we need at least $n + m$ qubits to perform the task.

The set of n qubits, the *input register*, encodes x, the set of m-qubits, the *output register*, represents the value $f(x)$. Having separate registers for input and output is standard practice in the classical theory of reversible computation.

In order to perform the calculation of $f(x)$, we should apply a suitable unitary transformation $\hat{U}_f$ to our set of $n + m$ qubits. The standard computational protocol defines the action of $\hat{U}_f$ on every computational basis state $|x\rangle_n |y\rangle_m$ of the $n + m$ qubits making up the input and output registers as follows:

$$\hat{U}_f |x\rangle_n |y\rangle_m = |x\rangle_n |y \oplus f(x)\rangle_m, \tag{3.25}$$

where $\oplus$ can be seen as a generalized XOR acting on the single bits belonging to the two strings of bits y and $f(x)$, respectively. Indeed, $\hat{U}_f |x\rangle_n |0\rangle_m = |x\rangle_n |f(x)\rangle_m$: by initializing the starting output register to $|0\rangle_m$, after the computation it represents the actual value $f(x)$.

3.4.1 Realistic Computation

The computation of $f(x)$ may require more than the $n + m$ qubit introduced in Sect. 3.4. In Fig. 3.14 it is sketched a more realistic quantum circuit to carry out the calculation of $f(x)$, where an additional register of r qubits and a unitary transformation $\hat{W}_f$ acting on $n + m + r$ qubit is used. As shown in the lower circuit of Fig. 3.14, the unitary $\hat{W}_f$ act as follows: the additional r-qubit state $|\Psi\rangle_r$ interact with the input register $|x\rangle_n$ through the unitary operation $\hat{V}_f$ obtaining the evolution:

$$\hat{V}_f |\Psi\rangle_r |x\rangle_n = |\Phi\rangle_{n+r-m} |f(x)\rangle_m. \tag{3.26}$$

3.5 Circuit Identities

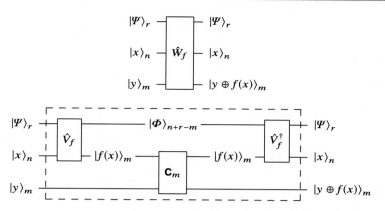

Fig. 3.14 Realistic view of the structure of a unitary transformation $\hat{W}_f$ to carry out the calculation of $f(x)$. See the text for details

Now, m CNOT gates perform, bit by bit, the addition modulo 2 with the state of the output register (the control qubits are in the state $|f(x)\rangle_m$):

$$\mathbf{C}_m |f(x)\rangle_m |y\rangle_m = |f(x)\rangle_m |y \oplus f(x)\rangle_m. \tag{3.27}$$

A final unitary $\hat{V}_f^\dagger$ is used to obtain the transformation:

$$\hat{V}_f^\dagger |\Phi\rangle_{n+r-m} |f(x)\rangle_m = |\Psi\rangle_r |x\rangle_n. \tag{3.28}$$

3.5 Circuit Identities

In Fig. 3.15 we report useful circuit identities that can be used to better understand the behavior of the quantum circuits described in the following sections. The reader can easily verify them.

Here we explicitly consider the identity (f), namely, $\mathbf{H} \otimes \mathbf{H} \mathbf{C}_{10} \mathbf{H} \otimes \mathbf{H}$. Since:

$$\mathbf{C}_{10} = \frac{1}{2}(\hat{\mathbb{I}} + \hat{\sigma}_z) \otimes \hat{\mathbb{I}} + \frac{1}{2}(\hat{\mathbb{I}} - \hat{\sigma}_z) \otimes \hat{\sigma}_x, \tag{3.29}$$

it is straightforward to verify that:

$$\mathbf{H} \otimes \mathbf{H} \mathbf{C}_{10} \mathbf{H} \otimes \mathbf{H} = \hat{\mathbb{I}} \otimes \frac{1}{2}(\hat{\mathbb{I}} + \hat{\sigma}_z) + \hat{\sigma}_x \otimes \frac{1}{2}(\hat{\mathbb{I}} - \hat{\sigma}_z) \equiv \mathbf{C}_{01}, \tag{3.30}$$

where we used the identities (b) and (c) of Fig. 3.15.

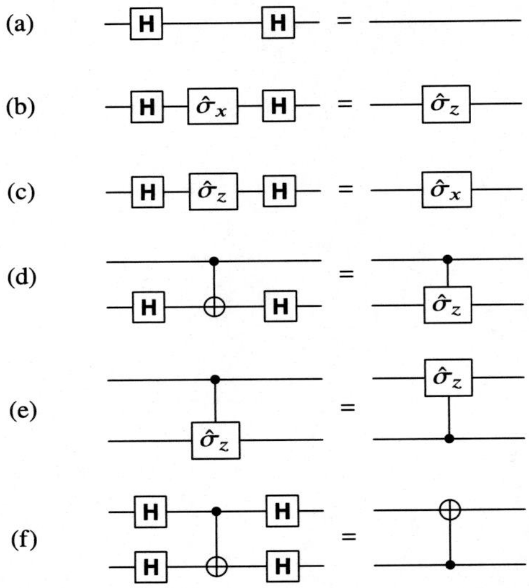

Fig. 3.15 Useful circuit identities

3.6 Introduction to Quantum Algorithms

As we have mentioned, a quantum algorithm involves two registers: the input register $|x\rangle_n$ and the output register $|y\rangle_m$. This is due to the reversibility of quantum operations: in general, a logical operation is not reversible, while the unitary operations indeed are. In this view, a quantum algorithm is similar to a classical reversible computation (of course, in the last case we cannot exploit the quantum features of qubits!).

We recall that the standard computational process that calculates $f(x)$, can be always represented as a suitable unitary operator $\hat{U}_f$ acting on the state $|x\rangle_n |y\rangle_m$, that is:

$$\hat{U}_f |x\rangle_n |y\rangle_m = |x\rangle_n |y \oplus f(x)\rangle_m. \tag{3.31}$$

Given a single qubit $|x\rangle$, $x = 0, 1$, the action of the Hadamard transformation **H** can be summarized as:

$$\mathbf{H}|x\rangle = \frac{|0\rangle + (-1)^x |1\rangle}{\sqrt{2}}, \tag{3.32a}$$

$$= \frac{1}{\sqrt{2}} \sum_{z=0,1} (-1)^{xz} |z\rangle. \tag{3.32b}$$

3.6 Introduction to Quantum Algorithms

Therefore, given an n-qubit state $|x\rangle_n$, with $0 \leq x < 2^n$ and $x = \sum_{k=0}^{n-1} x_k 2^k$ with $x_k \in \{0, 1\}$, we have:

$$\mathbf{H}^{\otimes n}|x\rangle_n = \left(\frac{|0\rangle + (-1)^{x_{n-1}}|1\rangle}{\sqrt{2}}\right) \otimes \cdots \otimes \left(\frac{|0\rangle + (-1)^{x_0}|1\rangle}{\sqrt{2}}\right), \quad (3.33a)$$

$$= \frac{1}{2^{n/2}} \sum_{z=0}^{2^n-1} (-1)^{x \cdot z} |z\rangle_n, \quad (3.33b)$$

where $x \cdot z = \oplus_{k=0}^{n-1} x_k z_k \pmod{2}$.

It is also useful to note that if $f(x) \in \{0, 1\}$, then:

$$\hat{U}_f (\hat{\mathbb{I}} \otimes \mathbf{H})|x\rangle|1\rangle = |x\rangle \frac{|f(x)\rangle - |1 \oplus f(x)\rangle}{\sqrt{2}}, \quad (3.34a)$$

$$= (-1)^{f(x)} |x\rangle \frac{|0\rangle - |1\rangle}{\sqrt{2}}. \quad (3.34b)$$

Therefore, the phase of the second qubit is transferred to the first one: this fact is called *phase kickback* and it is typical of the controlled operations that may affect the control qubit as well. We will see that the factor $(-1)^{f(x)}$ due to the phase kickback is extremely important for quantum algorithms creating computational advantages.

3.6.1 Deutsch Algorithm

The first pedagogical algorithm we consider has been proposed to solve the so-called Deutsch problem. Given a function $f : \{0, 1\} \to \{0, 1\}$, suppose we are not interested in the *particular* values $f(0)$ and $f(1)$, but rather in a *relational* information, that is whether $f(0) = f(1)$ or $f(0) \neq f(1)$. Form the classical point of view, the only way to solve this problem is to evaluate $f(x)$ twice. We are going to show that a quantum algorithm can tell us the answer by using just one evaluation of the function. The circuit implementing the algorithm is shown in Fig. 3.16.

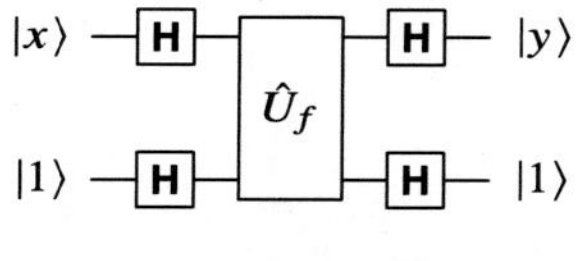

Fig. 3.16 Quantum circuit to solve the Deutsch problem (Deutsch algorithm): if $f(0) = f(1)$ then $|y\rangle = |x\rangle$, otherwise $|y\rangle = |\bar{x}\rangle$, thus measuring the input register after the query we can discriminate between the two possible kind of functions

The first step of the algorithm is to apply the Hadamard transformations to the qubit initial states:

$$\mathbf{H} \otimes \mathbf{H}|x\rangle|1\rangle = \sum_z \frac{(-1)^{xz}}{\sqrt{2}} |z\rangle \left(\frac{|0\rangle - |1\rangle}{\sqrt{2}}\right) \quad (3.35)$$

where $z = 0, 1$. Now we apply $\hat{U}_f$ (note the phase kickback):

$$\hat{U}_f \sum_z \frac{(-1)^{xz}}{\sqrt{2}} |z\rangle \left(\frac{|0\rangle - |1\rangle}{\sqrt{2}}\right) = \sum_z \frac{(-1)^{xz+f(z)}}{\sqrt{2}} |z\rangle \left(\frac{|0\rangle - |1\rangle}{\sqrt{2}}\right). \quad (3.36)$$

Finally, we use again the Hadamard transformations, obtaining the following whole evolution:

$$(\mathbf{H} \otimes \mathbf{H}) \hat{U}_f (\mathbf{H} \otimes \mathbf{H})|x\rangle|1\rangle = \sum_s c_f(x, s)|s\rangle|1\rangle, \quad (3.37)$$

where $s = 0, 1$, and we introduced the coefficients:

$$c_f(x, s) = \frac{1}{2}(-1)^{f(0)} \left[1 + (-1)^{x+s}(-1)^{f(1)-f(0)}\right]. \quad (3.38)$$

We note that this latter step makes the different amplitudes interfere: this is what allows us achieving the final result of the computation. "Quantum interference" is another key ingredient that quantum computation can exploit to overcome the limits of the classical one.

Let us come back to the Deutsch algorithm. After the last Hadamard transformation the output register has been left unchanged, since it is still in the state $|1\rangle$, while the input register has undergone the evolution:

$$|x\rangle \rightarrow \sum_s c_f(x, s)|s\rangle. \quad (3.39)$$

If $f(x) \in \{0, 1\}$, then it is straightforward to verify that:

- if $f(1) = f(0) \Rightarrow |c_f(x, x)|^2 = 1$ and $|c_f(x, \bar{x})|^2 = 0$,
- if $f(1) \neq f(0) \Rightarrow |c_f(x, x)|^2 = 0$ and $|c_f(x, \bar{x})|^2 = 1$,

therefore, if a measurement on the input register gives a result $|x\rangle$, we can conclude that $f(1) = f(0)$, if it leads to $|\bar{x}\rangle$, we have $f(1) \neq f(0)$. This happens after a single query of $\hat{U}_f$. Note that we do not know the actual value of $f(1)$ and $f(0)$: this is a typical quantum tradeoff that scarifies particular information to acquire relational information.

3.6.2 Deutsch–Jozsa Algorithm

Now our function is

$$f : \{0, 1\}^{\otimes n} \to \{0, 1\}, \qquad (3.40)$$

that is $f(x) \in \{0, 1\}$ but $0 \le x < 2^n$. We assume to know that f can only have the following two mutual exclusive properties:

- or f is *constant*: $f(x) = f(0), \forall x$;
- or f is *balanced*: $f(x) = 1$, for half of the possible 2^n values of x, otherwise $f(x) = 0$.

The problem is to decide whether f is balanced or not.

In the best case a *deterministic* classical computer may solve the problem with just two queries [if $f(0) \ne f(1)$ then f is indeed balanced]. However in the worst case it could happen that the first $2^n/2 = 2^{n-1}$ queries give the same output, then we need one more query to answer the problem: if we have still the same result f is constant, otherwise it is balanced.

A classical *randomized* algorithm can indeed do better. This algorithm randomly chooses $m \le 2^{n-1}$ values of x, obtaining the set $\{x^{(1)}, \ldots x^{(m)}\}$, and compare the value $f(x^{(k)})$ with that of $f(x^{(1)})$, $1 < k \le m$. Given a balanced f and the value $f(x^{(1)})$, the probability that $f(x^{(k)}) = f(x^{(1)})$ is $1/2$. Therefore the *probability of failure*, that is the probability that $f(x^{(1)}) = f(x^{(k)})$, $\forall k$, is:

$$p_{\text{fail}}(m) = \underbrace{\frac{1}{2} \times \frac{1}{2} \times \ldots \times \frac{1}{2}}_{(m-1)\text{-times}} = \frac{1}{2^{m-1}}, \qquad (3.41)$$

where we consider only $m - 1$ values of x because the first one is used as control. We thus obtain that after m queries, the probability of success, i.e., we find that f is balanced, is $p_{\text{succ}}(m) = 1 - p_{\text{fail}}(m)$.

In Fig. 3.17 we can see the quantum circuit to solve the Deutsch–Jozsa problem. The input states is the $n + 1$ qubit state $|0\rangle_n|1\rangle$, and, after the application of the

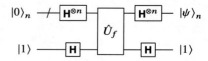

Fig. 3.17 Quantum circuit to solve the Deutsch–Jozsa problem

Hadamard transformations and the query of $\hat{U}_f$, we have:

$$\hat{U}_f(\mathbf{H}^{\otimes n} \otimes \mathbf{H})|0\rangle_n|1\rangle = \hat{U}_f \left(\frac{1}{2^{n/2}} \sum_{x=0}^{2^n-1} |x\rangle_n \right) \left(\frac{|0\rangle - |1\rangle}{\sqrt{2}} \right)$$

$$= \frac{1}{2^{n/2}} \sum_{x=0}^{2^n-1} (-1)^{f(x)} |x\rangle_n \left(\frac{|0\rangle - |1\rangle}{\sqrt{2}} \right). \quad (3.42)$$

Now we should apply the Hadamard transformations to get the interference among the amplitudes (note the phase kickback):

$$(\mathbf{H}^{\otimes n} \otimes \mathbf{H}) \hat{U}_f (\mathbf{H}^{\otimes n} \otimes \mathbf{H})|0\rangle_n|1\rangle = \frac{1}{2^n} \sum_{z=0}^{2^n-1} \sum_{x=0}^{2^n-1} (-1)^{z \cdot x + f(x)} |z\rangle_n |1\rangle \quad (3.43a)$$

$$= |\psi\rangle_n |1\rangle, \quad (3.43b)$$

where:

$$|\psi\rangle_n = \sum_{z=0}^{2^n-1} c_f(z) |z\rangle_n, \quad (3.44)$$

with

$$c_f(z) = \frac{1}{2^n} \sum_{x=0}^{2^n-1} (-1)^{z \cdot x + f(x)}. \quad (3.45)$$

We can focus on:

$$c_f(0) = \frac{1}{2^n} \sum_{x=0}^{2^n-1} (-1)^{f(x)}. \quad (3.46)$$

On the one hand, if $f(x)$ is *constant*, namely $f(x) = f(0)$, $\forall x$, we have $c_f(0) = (-1)^{f(0)}$, and, since $|\psi\rangle_n$ should be normalized, i.e., $\sum_z |c_f(z)|^2 = 1$, we obtain $c_f(z) = 0$, $\forall z \neq 0$. On the other hand, if $f(x)$ is *balanced* we get $c_f(0) = 0$, since the sum in Eq. (3.46) contains 2^{n-1} times the value " $+1$" and 2^{n-1} times the value " -1" and, thus, the corresponding state $|\psi\rangle_n$ does not contain $|0\rangle_n$.

3.6 Introduction to Quantum Algorithms

Summarizing, the Deutsch–Jozsa algorithm leads to the following evolution of the n-qubit input register:

$$|0\rangle_n \to \begin{cases} |0\rangle_n & \text{if } f \text{ is constant,} \\ \sum_{z=1}^{2^n-1} c_f(z)|z\rangle_n & \text{if } f \text{ is balanced,} \end{cases} \quad (3.47)$$

therefore, just after a single call of U_f, a measurement of the evolved state of the input register allows us to decide if f is constant (we obtain $|0\rangle_n$) or balanced (in this last case we have $|x\rangle_n, x \neq 0$).

It is worth noting that: (1) there is not any known practical application of this kind of algorithm; (2) the method used to evaluate $f(x)$ is different in the classical and in the quantum case; (3) the probabilistic algorithms can find the solution of the Deutsch–Jozsa problem with high probability just after few (random) evaluations of $f(x)$.

3.6.3 Bernstein–Vazirani Algorithm

Let a be an unknown integer number, $0 \leq a < 2^n$ and consider the function:

$$f(x) = a \cdot x \equiv a_0 x_0 \oplus \cdots \oplus a_{n-1} x_{n-1}. \quad (3.48)$$

The problem is to find the unknown a given a subroutine that evaluates $f(x)$ for an integer $0 \leq x < 2^n$. Classically the only way to solve the problem is to evaluate $f(2^m) \equiv a_m$ for $m = 0, 1, \ldots, n-1$, which, thus, requires n evaluations of $f(x)$.

Figure 3.18 shows the quantum-circuit representation of the Bernstein–Vazirani problem. The input register encodes the n-qubit state $|x\rangle_n = |x_{n-1}\rangle \otimes \ldots \otimes |x_0\rangle$ while the output register, which is initialized to $|0\rangle$, after the evolution through the unitary operator $\hat{U}_f$ associated with the function defined in Eq. (3.48), becomes $|a \cdot x\rangle$.

The quantum circuit we need to solve the present problem with just one call of $\hat{U}_f$ is the same of the Deutsch–Jozsa problem (see Fig. 3.17). Since, now, the action of f is given in Eq. (3.48), the coefficients of the state in Eq. (3.44) read:

$$c_f(z) = \frac{1}{2^n} \sum_{x=0}^{2^n-1} (-1)^{z \cdot x + a \cdot x}, \quad (3.49a)$$

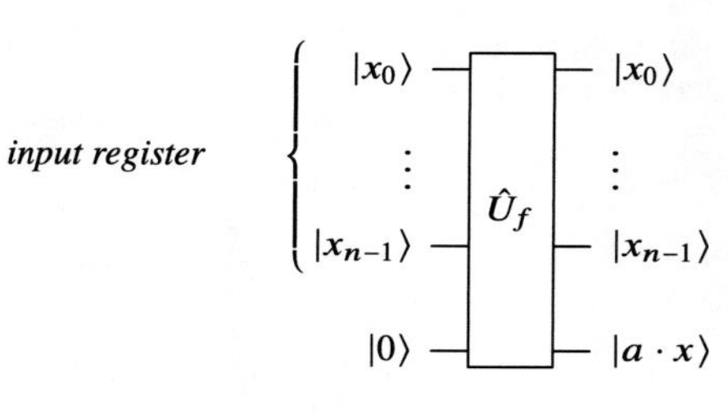

Fig. 3.18 The Bernstein–Vazirani problem

$$= \frac{1}{2^n} \prod_{k=0}^{n-1} \sum_{x_k=0,1} (-1)^{(z_k+a_k)x_k}, \qquad (3.49b)$$

$$= \frac{1}{2^n} \prod_{k=0}^{n-1} \left[1 + (-1)^{z_k+a_k}\right]. \qquad (3.49c)$$

Form the last equality we conclude that if there exists k such that $z_k \neq a_k$, then $c_f(z) = 0$. Therefore we have:

$$c_f(z) \to \begin{cases} 0 \text{ if } z \neq a, \\ 1 \text{ if } z = a, \end{cases} \qquad (3.50)$$

that is, the evolution of the input register can be summarized as:

$$|0\rangle_n \to |a\rangle_n, \qquad (3.51)$$

and the measurement of the evolved input register in the computational basis directly gives the unknown value of a.

A further investigation of the quantum circuit implementing $\hat{U}_f$ may explain the mechanism underling the Bernstein–Vazirani algorithm. In particular, in Fig. 3.19 we illustrate the quantum circuit, based on CNOT gates, used to calculate $f(x) = a \cdot x$ in the case of $n = 4$. It is clear that the value y of the output register is flipped only if a_k and x_k are both equal to 1, since the CNOT taking $|x_k\rangle$ as control bit is present in the circuit only if $a_k = 1$.

As depicted in Fig. 3.20 (top), the solution of the problem consists in the application of the Hadamard transformation before and after the unitary U_f. But since $\mathbf{C}_{hk} = \mathbf{H}_h \mathbf{H}_k \mathbf{C}_{kh} \mathbf{H}_h \mathbf{H}_k$ (see problem 1.7 and Sect. 3.5), the resulting circuit is equivalent to the one depicted in Fig. 3.20 (bottom): it is now straightforward to see that by taking $|0\rangle_n$ and $|1\rangle$ as initial states of the input and output registers, respectively, one has $|0\rangle_n \to |a\rangle_n$

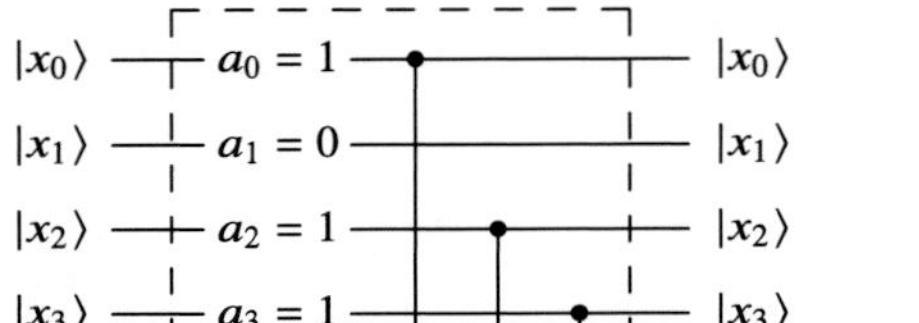

Fig. 3.19 The quantum circuit to implement the Bernstein–Vazirani problem for $a = 1101$: if $a_k = 1$ then the bit k-th bit acts as control for the NOT operation on to the output register (lowest wire)

Fig. 3.20 (Top) Quantum solution of the Bernstein–Vazirani problem. (Bottom) The equivalent quantum circuit to solve the problem for $a = 1101$: we used the identity $\mathbf{C}_{hk} = \mathbf{H}_h \mathbf{H}_k \mathbf{C}_{kh} \mathbf{H}_h \mathbf{H}_k$

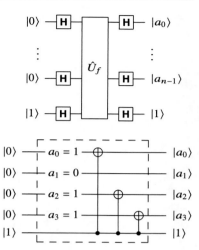

3.7 Classical Logic with Quantum Computers

We have underlined that classical logic is intrinsically irreversible, due to the loss of information during the computation. However, classical computers, based on this irreversible framework, are of course extremely useful tools to deal with a plethora of problems. Being irreversible, classical gates such as AND and OR cannot be *directly* implemented in the quantum scenario, but one can resort to particular reversible quantum gates able to perform the action of any classical gate and, in turn, computation, in a reversible fashion.

3.7.1 The Toffoli Gate

Any arithmetical operation can be built up on a reversible classical computer out of three-bit controlled-controlled-NOT (CCNOT) gates called Toffoli gates.

The Toffoli gate, represented by the unitary operator **T**, acts on a 3-bit state as follows:

$$\mathbf{T}|x\rangle|y\rangle|z\rangle = |x\rangle|y\rangle|z \oplus xy\rangle, \tag{3.52}$$

where xy corresponds to the arithmetical product between the values x and y. The action of the Toffoli gate onto the computational basis is summarized in Table 3.1. As one can see, **T** leaves unchanged the third bit, unless the state of the control bits are in the state $|1\rangle|1\rangle$, in this case the value of the target bit is flipped (see the last two lines of the table). Of course **T** is reversible and its action on the computational basis is a permutation. The quantum circuit for the Toffoli gate is shown in Fig. 3.21.

Table 3.1 The action of the Toffoli gate

$\|x\rangle\|y\rangle\|z\rangle$	$\mathbf{T}\|x\rangle\|y\rangle\|z\rangle$
$\|0\rangle\|0\rangle\|0\rangle$	$\|0\rangle\|0\rangle\|0\rangle$
$\|0\rangle\|0\rangle\|1\rangle$	$\|0\rangle\|0\rangle\|1\rangle$
$\|0\rangle\|1\rangle\|0\rangle$	$\|0\rangle\|1\rangle\|0\rangle$
$\|0\rangle\|1\rangle\|1\rangle$	$\|0\rangle\|1\rangle\|1\rangle$
$\|1\rangle\|0\rangle\|0\rangle$	$\|1\rangle\|0\rangle\|0\rangle$
$\|1\rangle\|0\rangle\|1\rangle$	$\|1\rangle\|0\rangle\|1\rangle$
$\|1\rangle\|1\rangle\|0\rangle$	$\|1\rangle\|1\rangle\|1\rangle$
$\|1\rangle\|1\rangle\|1\rangle$	$\|1\rangle\|1\rangle\|0\rangle$

Fig. 3.21 Quantum circuit for the Toffoli gate

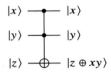

As we mentioned in Chap. 1, all the logical and, thus, arithmetical operations can be built up out of AND and NOT. By using the Toffoli gate one can calculate the logical AND of two bits, which corresponds to the product of their values, and the NOT, namely:

$$\text{AND} \rightarrow \mathbf{T}|x\rangle|y\rangle|0\rangle = |x\rangle|y\rangle|xy\rangle \equiv |x\rangle|y\rangle|x \wedge y\rangle, \tag{3.53a}$$

$$\text{NOT} \rightarrow \mathbf{T}|1\rangle|1\rangle|z\rangle = |1\rangle|y\rangle|z \oplus 1\rangle \equiv |1\rangle|1\rangle|\bar{z}\rangle, \tag{3.53b}$$

respectively. We have just demonstrated the universality of the Toffoli gate. Furthermore, we have:

$$\text{XOR} \rightarrow \mathbf{T}|1\rangle|y\rangle|z\rangle = |1\rangle|y\rangle|z \oplus y\rangle, \tag{3.54a}$$

$$\text{NAND} \rightarrow \mathbf{T}|x\rangle|y\rangle|1\rangle = |x\rangle|y\rangle|\overline{x \wedge y}\rangle, \tag{3.54b}$$

We can conclude that it is possible to do any computation reversibly.

We have seen the importance of the Toffoli gate. However, this gate cannot be realized by means of one- or two-bit classical gates. Fortunately, there exist quantum gates! In Fig. 3.22 is depicted a quantum circuit that acts as a controlled-controlled-$\hat{U}^2$ gate (a CC-$\hat{U}^2$), where $\hat{U}$ is a unitary operator ($\hat{U}\hat{U}^\dagger = \hat{U}^\dagger \hat{U} = \hat{\mathbb{I}}$), that involves only CNOT and controlled-$\hat{U}$ gates (C-$\hat{U}$). The reader can easily verify that the circuit applies the $\hat{U}^2$ operator to the state $|z\rangle$ of the output register only if the two-bit input register is $|x\rangle|y\rangle = |1\rangle|1\rangle$, namely:

$$|x\rangle|y\rangle|z\rangle \rightarrow \hat{\mathbb{I}} \otimes \hat{\mathbb{I}} \otimes \left[\hat{U}^y(\hat{U}^\dagger)^{x \oplus y}\hat{U}^x\right]|x\rangle|y\rangle|z\rangle. \tag{3.55}$$

3.7 Classical Logic with Quantum Computers

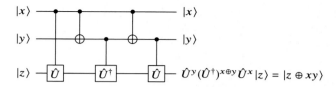

Fig. 3.22 Quantum circuit acting as a CC-$\hat{U}^2$ gate based on CNOT and C-$\hat{U}$ gates. If we choose $\hat{U} = \sqrt{\hat{\sigma}_x}$ (the square root of NOT) we can reproduce the effect of the Toffoli gate

Table 3.2 The action of the Fredkin gate

$\|x\rangle\|y\rangle\|z\rangle$	$F\|x\rangle\|y\rangle\|z\rangle$
$\|0\rangle\|0\rangle\|0\rangle$	$\|0\rangle\|0\rangle\|0\rangle$
$\|0\rangle\|0\rangle\|1\rangle$	$\|0\rangle\|0\rangle\|1\rangle$
$\|0\rangle\|1\rangle\|0\rangle$	$\|0\rangle\|1\rangle\|0\rangle$
$\|0\rangle\|1\rangle\|1\rangle$	$\|0\rangle\|1\rangle\|1\rangle$
$\|1\rangle\|0\rangle\|0\rangle$	$\|1\rangle\|0\rangle\|0\rangle$
$\|1\rangle\|0\rangle\|1\rangle$	$\|1\rangle\|1\rangle\|0\rangle$
$\|1\rangle\|1\rangle\|0\rangle$	$\|1\rangle\|0\rangle\|1\rangle$
$\|1\rangle\|1\rangle\|1\rangle$	$\|1\rangle\|1\rangle\|1\rangle$

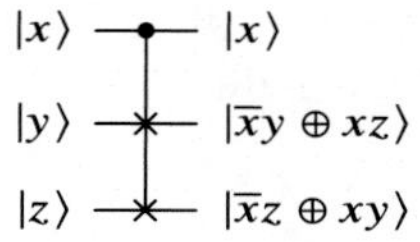

Fig. 3.23 Quantum circuit for the Fredkin gate

If we now introduce the unitary operator:

$$\hat{U} = \sqrt{\hat{\sigma}_x} \rightarrow \frac{1}{1+i}\begin{pmatrix} 1 & i \\ i & 1 \end{pmatrix} \quad (3.56)$$

(square root of NOT)

such that $\hat{U}^2 = \sqrt{\hat{\sigma}_x}\sqrt{\hat{\sigma}_x} = \hat{\sigma}_x$, then the ccNOT can be obtained with the quantum circuit of Fig. 3.22. Note that $\sqrt{\hat{\sigma}_x}$ does not exist as a classical gate, but it exists as quantum gate, since:

$$\sqrt{\hat{\sigma}_x}|0\rangle = \frac{|0\rangle + i|1\rangle}{1+i}, \quad \text{and} \quad \sqrt{\hat{\sigma}_x}|1\rangle = \frac{i|0\rangle + |1\rangle}{1+i}. \quad (3.57)$$

3.7.2 The Fredkin Gate

The Fredkin gate is another three-bit gate which can be used to build a universal set of gates. This gate has one control bit and two targets: when the control bit is 1 the targets are swapped, otherwise they are left unchanged. The action of the Fredkin gate, represented by the unitary operator **F**, is summarized in Table 3.2, whereas we show the corresponding quantum circuit in Fig. 3.23.

By suitably setting the input bits it is possible to implement any logical operation. For instance we have:

$$\text{AND} \to \mathbf{F}|x\rangle|y\rangle|0\rangle = |x\rangle|\overline{x} \wedge y\rangle|x \wedge y\rangle, \tag{3.58a}$$

$$\text{NOT} \to \mathbf{F}|x\rangle|0\rangle|1\rangle = |x\rangle|x\rangle|\overline{x}\rangle, \tag{3.58b}$$

therefore the Fredkin gate is universal. Note that in the last case we implemented both the COPY and the NOT operations at the same time.

3.8 Universal Quantum Gates

Universal quantum computation can be performed by means of any entangling interaction together with local unitaries, that is any unitary operator acting on a system of qubits can be reduced to a product of operators which entangle two qubits or act locally on a single qubit. In order to prove this claim, in this section first we show that any unitary acting on d levels can be decomposed in a product of unitaries acting at most on only two levels; then we prove that any two-level unitary can be implemented by means of CNOT and single-qubit gates, the first ones being the entangling gates whereas the others perform local operations on the qubits.

3.8.1 Universality of Two-Level Unitaries

Any unitary $\mathbf{U}$ acting on d levels can be decomposed in a product of two-level unitaries corresponding to suitable unitary transformations $\mathbf{G}(h, k)$, with $h < k$, acting on the levels h and k associated with the states $|h\rangle$ and $|k\rangle$ of the computational basis, $h, k \in \{1, \ldots, d\}$. The transformation $\mathbf{G}(h, k)$ can be seen as a $d \times d$ matrix whose elements are:

$$\begin{cases} g_{hh}(h, k) = c^*, & g_{kk}(h, k) = -c \\ g_{hk}(h, k) = s^*, & g_{kh}(h, k) = s \\ g_{pp}(h, k) = 1 & \text{if } p \neq h, k \\ g_{pq}(h, k) = 0 & \text{otherwise,} \end{cases} \tag{3.59}$$

where c and s are complex numbers (see Fig. 3.24). We will use this kind of transform to reduce $\mathbf{U}$ to a product of two-level unitaries, since when $\mathbf{G}(h, k)$ is applied to $\mathbf{U}$ it acts only on the two levels h and k.

In order to show how the method works, we focus on the simple three-level example (that is $d = 3$) and the reader will see that the extension to larger dimensions is straightforward. The basic idea is to exploit the transformations defined above to put equal to zero the matrix elements u_{p1} of the first column of

3.8 Universal Quantum Gates

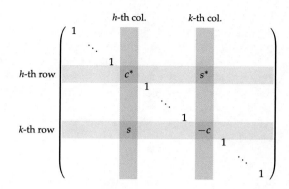

Fig. 3.24 Matrix representation of the unitary transformation $\mathbf{G}(h, k)$

$\mathbf{U}$, but u_{11} which is left equal to 1. We proceed as follows. If we choose $c = u_{11}/\mu$ and $s = u_{21}/\mu$, where $\mu = \sqrt{|u_{11}|^2 + |u_{21}|^2}$, we have (for the sake of clarity we focus on the first column only and use the symbol "$*$" for all the others):

$$\mathbf{G}(1, 2)\, \mathbf{U} = \begin{pmatrix} \mu & * & * \\ 0 & * & * \\ u_{31} & * & * \end{pmatrix} \tag{3.60}$$

It is clear that the action is to perform the transformation $u_{11} \to \mu$ and $u_{21} \to 0$ on the first column (as expected the element u_{31} is left unchanged). Now, acting in a similar way, we apply $\mathbf{G}(1, 3)$ by setting $c = \mu/\mu'$ and $s = u_{31}/\mu'$, where $\mu' = \sqrt{\mu^2 + |u_{31}|^2} = 1$ (since $\mathbf{U}$ is unitary we have $\sum_p |u_{p\tilde{q}}|^2 = \sum_q |u_{\tilde{p}q}|^2 = 1$, $\forall \tilde{p}, \tilde{q}$) and obtain:

$$\mathbf{G}(1, 3)\, \mathbf{G}(1, 2)\, \mathbf{U} = \begin{pmatrix} 1 & 0 & 0 \\ 0 & * & * \\ 0 & * & * \end{pmatrix} \equiv V_{2,3}. \tag{3.61}$$

Note that $V_{2,3}$ is a unitary operator acting on the two levels 2 and 3. Form Eq. (3.61) directly follows that $\mathbf{U}$ can be written as the product of the following two-levels unitary operators:

$$\mathbf{U} = V_{1,2} V_{1,3} V_{2,3}, \tag{3.62}$$

where $V_{1,2}^\dagger = G(1, 2)$ and $V_{1,3}^\dagger = G(1, 3)$.

For a generic dimension d, one should repeat the previous procedure to the other columns to obtain a final matrix of the form (only the non-zero elements are reported):

$$\begin{pmatrix} \mathbb{I}_{d-2} & & \\ & * & * \\ & * & * \end{pmatrix}, \tag{3.63}$$

which acts only on the last two levels. It is also possible to show that the number of needed two-level unitaries is at most $d(d-1)/2$. This concludes the proof of the universality of two-level unitaries.

It is worth noting that a system of n qubits encodes $d = 2^n$ levels, i.e., $|x\rangle_n = |x_{n-1}\rangle \ldots |x_0\rangle$ with $x \in \{0, 1, \ldots, 2^n - 1\}$ and $x_k \in \{0, 1\}$, and, thus, a two-level unitary may also couples two levels belonging to different qubits and the number of needed two-level unitaries is at most $2^n(2^n - 1)/2 \sim O(4^n)$. In the next section we will show that any two-level unitary can be implemented by using single-qubit and CNOT gates.

3.8.2 Universality of Single-Qubit and CNOT Gates

For the sake of clarity (and simplicity!) here we focus on a system of three qubits, that is $d = 2^3 = 8$, spanned by the computational basis, that correspond to the eight levels we mentioned in the previous sections:

$$|0\rangle_3 = |0\rangle|0\rangle|0\rangle, \quad |1\rangle_3 = |0\rangle|0\rangle|1\rangle, \tag{3.64a}$$

$$|2\rangle_3 = |0\rangle|1\rangle|0\rangle, \quad |3\rangle_3 = |0\rangle|1\rangle|1\rangle, \tag{3.64b}$$

$$|4\rangle_3 = |1\rangle|0\rangle|0\rangle, \quad |5\rangle_3 = |1\rangle|0\rangle|1\rangle, \tag{3.64c}$$

$$|6\rangle_3 = |1\rangle|1\rangle|0\rangle, \quad |7\rangle_3 = |1\rangle|1\rangle|1\rangle. \tag{3.64d}$$

We consider a unitary matrix 8×8 matrix **U** acting only on the two levels $|x\rangle_3$ and $|y\rangle_3$, $x < y$, which has the following entries (see Fig. 3.25):

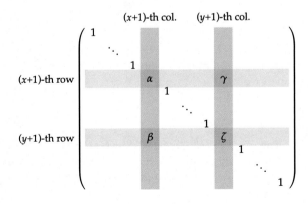

Fig. 3.25 Matrix representation of the unitary **U**

3.8 Universal Quantum Gates

$$\begin{cases} u_{x+1\,x+1} = \alpha, \; u_{x+1\,y+1} = \gamma, \\ u_{y+1\,x+1} = \beta, \; u_{y+1\,y+1} = \zeta, \\ u_{pp} = 1 \qquad \text{if } p \neq x+1, y+1 \\ u_{pq} = 0 \qquad \text{otherwise,} \end{cases} \quad (3.65)$$

which act as follows:

$$\begin{cases} \mathbf{U}|x\rangle_3 = \alpha|x\rangle_3 + \beta|y\rangle_3, \\ \mathbf{U}|y\rangle_3 = \gamma|x\rangle_3 + \zeta|y\rangle_3, \\ \mathbf{U}|p\rangle_3 = |p\rangle_3 \qquad \text{if } p \neq x, y. \end{cases} \quad (3.66)$$

We look for a quantum circuit implementing $\mathbf{U}$, built from single-qubit and CNOT gates. The trick is to use the so-called *Gray codes*: a Gray code connecting two binary number x and y is a sequence of m binary numbers $\{b_1, b_2, \ldots b_m\}$ starting from $b_1 = x$ and arriving at $b_m = y$, such that adjacent numbers differ in exactly one bit.

As an example, we consider $x = 0$ and $y = 7$, that is, $|x\rangle_3 = |0\rangle|0\rangle|0\rangle$ and $|y\rangle_3 = |1\rangle|1\rangle|1\rangle$. The Gray code connecting the two numbers (or, equivalently, states), is (for the sake of clarity we added a subscript to identify the three qubits):

$$|0\rangle_3 = |0\rangle_A|0\rangle_B|0\rangle_C \quad (3.67a)$$

$$\rightarrow |0\rangle_A|0\rangle_B|1\rangle_C \quad (3.67b)$$

$$\rightarrow |0\rangle_A|1\rangle_B|1\rangle_C \quad (3.67c)$$

$$\rightarrow |1\rangle_A|1\rangle_B|1\rangle_C = |7\rangle_3. \quad (3.67d)$$

To pass from one element to the adjacent one, we can easily use CCNOT gates (NOT gates controlled by two qubits), which have as target the only different bit (here qubit A) and uses the values of the others as control (here B and C). We can use this strategy to connect step-wise $|x\rangle_3 = |0\rangle_A|0\rangle_B|0\rangle_C$ to $|0\rangle_A|1\rangle_B|1\rangle_C$, the element just before $|y\rangle_3 = |1\rangle_A|1\rangle_B|1\rangle_C$, and then act with a CC-$\mathbf{U}_{xy}$ *on the first qubit*, namely, the unitary operator $\mathbf{U}_{xy}$ defined as the 2×2 matrix:

$$\mathbf{U}_{xy} = \begin{pmatrix} \alpha & \gamma \\ \beta & \zeta \end{pmatrix} \quad (3.68)$$

is applied to qubit A only if the others are in the state $|1\rangle_B|1\rangle_C$. Finally, we apply the same CCNOT gates, but in the reversed order. The overall circuit is shown in

Fig. 3.26 Quantum circuit implementing a unitary $\mathbf{U}_{xy}$ acting on the two levels (or states) $|x\rangle_3 = |0\rangle_A|0\rangle_B|0\rangle_C$ and $|y\rangle_3 = |1\rangle_A|1\rangle_B|1\rangle_C$

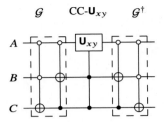

Fig. 3.26. The action of the first two gates can be summarized an operator $\mathcal{G}$ such that:

$$\mathcal{G}|0\rangle_3 = |0\rangle_A|1\rangle_B|1\rangle_C, \quad (3.69)$$

$$\mathcal{G}|7\rangle_3 = |1\rangle_A|1\rangle_B|1\rangle_C, \quad (3.70)$$

whereas when it is applied to the other states $|p\rangle_3$, with $p \neq 0, 7$, the final state of qubit B and C is $|0\rangle_B|0\rangle_C$ or $|b\rangle_B|c\rangle_C$, with $b \neq c$. Therefore, CC-$\mathbf{U}_{xy}$ can act non trivially only on the two states $|0\rangle_A|1\rangle_B|1\rangle_C$ and $|1\rangle_A|1\rangle_B|1\rangle_C$, that is:

$$\begin{aligned} \text{CC-}\mathbf{U}_{xy}\,\mathcal{G}|0\rangle_3 &= \text{CC-}\mathbf{U}_{xy}|0\rangle_A|1\rangle_B|1\rangle_C \\ &= \alpha|0\rangle_A|1\rangle_B|1\rangle_C + \beta|1\rangle_A|1\rangle_B|1\rangle_C, \end{aligned} \quad (3.71)$$

$$\begin{aligned} \text{CC-}\mathbf{U}_{xy}\,\mathcal{G}|7\rangle_3 &= \text{CC-}\mathbf{U}_{xy}|1\rangle_A|1\rangle_B|1\rangle_C \\ &= \gamma|0\rangle_A|1\rangle_B|1\rangle_C + \zeta|1\rangle_A|1\rangle_B|1\rangle_C, \end{aligned} \quad (3.72)$$

$$\text{CC-}\mathbf{U}_{xy}\,\mathcal{G}|p\rangle_3 = \mathcal{G}|p\rangle_3, \quad \text{if } |p\rangle_3 \neq |0\rangle_3, |7\rangle_3, \quad (3.73)$$

since $\mathbf{U}_{xy}|0\rangle_A = \alpha|0\rangle_A + \beta|1\rangle_A$ and $\mathbf{U}_{xy}|1\rangle_A = \gamma|0\rangle_A + \zeta|1\rangle_A$. Finally, the action of $\mathcal{G}^\dagger$ leads to the wanted transformation:

$$\begin{cases} \mathcal{G}^\dagger \text{ CC-}\mathbf{U}_{xy}\,\mathcal{G}|0\rangle_3 = \alpha|0\rangle_3 + \beta|1\rangle_3, \\ \mathcal{G}^\dagger \text{ CC-}\mathbf{U}_{xy}\,\mathcal{G}|7\rangle_3 = \gamma|0\rangle_3 + \zeta|1\rangle_3, \\ \mathcal{G}^\dagger \text{ CC-}\mathbf{U}_{xy}\,\mathcal{G}|p\rangle_3 = |p\rangle_3, \quad \text{if } |p\rangle_3 \neq |0\rangle_3, |7\rangle_3, \end{cases} \quad (3.74)$$

that is $\mathcal{G}^\dagger \text{ CC-}\mathbf{U}_{xy}\,\mathcal{G} \equiv \mathbf{U}$, as one can also see from Eq. (3.66) with $x = 0$ and $y = 7$. We note that the controlled-controlled gates acting on the three qubits can be implemented using only CNOT and single-qubit gates (see Fig. 3.22).

In the general case of n qubits, where we have 2^n levels, one can extend the previous protocol based on Gray codes. If $|g_1\rangle_n, |g_2\rangle_n, \ldots, |g_m\rangle_n$ are the m elements of the Gray code connecting $|g_1\rangle_n = |x\rangle_n$ and $|g_m\rangle_n = |y\rangle_n$, we can always find a code such that $m \leq n + 1$ (in fact $|x\rangle_n$ and $|y\rangle_n$ can differ in at most n locations). By using controlled gates we pass from $|g_1\rangle_n$ to $|g_{m-1}\rangle_n$, then

3.8 Universal Quantum Gates

we apply the controlled $\mathbf{U}_{xy}$ to the qubit located at the single bit where $|g_{m-1}\rangle_n$ and $|g_m\rangle_n$ differ, finally we undo the transformations of the first stage. Concerning the implementation, one can easily extend the scheme presented in Fig. 3.22 to system involving more than three qubits by suitably adding other CNOT and single-qubit gates. Therefore, thanks to the result obtained in the previous section (the universality of two-level unitaries), we have eventually proved also that CNOT and single-qubit gates are universal.

We note that the implementation of a unitary operation acting on n qubit requires a quantum circuit containing $O(n^2 4^n)$ single-qubit and CNOT gates. In fact, a two-level unitary requires $O(n^2)$ gates ($\mathcal{G}$ and the CC-$\mathbf{U}_{xy}$ both need $O(n)$ CNOT and single-qubit gates) and an arbitrary unitary requires $O(4^n)$ gates, as shown in the previous section. As a matter of fact, the approach followed in this universality construction does not provide efficient quantum circuits... In order to find fast algorithms one should use a different approach.

3.8.3 Set of Universal Quantum Gates

We have shown that single-qubit and CNOT gates can be used to perform universal quantum computation. However, there isn't a straightforward method to implement all these to be resistant to errors. On the other hand, it is possible to find a discrete set of gates to *approximate* any unitary operation. The *standard set* of universal gates consists of the Hadamard, phase, CNOT and T (or $\frac{\pi}{8}$ gate), introduced in Sect. 3.1. In reality Hadamard, CNOT and T gates may approximate a quantum circuit, but the presence of the phase gate S, that, formally, acts as T^2, is justified since it allows to approximate the circuit fault-tolerantly. In fact, during a computation the involved gates should be implemented assuming that the input qubits may be affected by errors. Therefore, as we will see in Chap. 8, the quantum gates should operate on multiple qubits generated from a suitable quantum error-correcting code. Also in the presence of faulty gates one can perform arbitrarily good quantum computation if the gate error probability is below a given threshold. In this scenario, the phase gate, as self-standing gate, turns out to be extremely useful.

3.8.4 Approximation of Single-Qubit Gates

First of all, we recall that a single-qubit gate described by the unitary operator $\hat{U}$ can be decomposed as (we drop the possible overall phase):

$$\hat{U} \rightarrow U = e^{i\theta \, \boldsymbol{v} \cdot \boldsymbol{\sigma}} = \mathcal{R}_{\boldsymbol{n}}(\beta) \mathcal{R}_{\boldsymbol{m}}(\gamma) \mathcal{R}_{\boldsymbol{n}}(\delta), \qquad (3.75)$$

where $\boldsymbol{n}$ and $\boldsymbol{m}$ are orthogonal (unit) vectors and the angles β, γ and δ depends on $\boldsymbol{v}$ and θ. We also observe that if α is and *irrational* multiple of 2π, then the angles $\kappa\alpha \pmod{2\pi}$, $\kappa \in \mathbb{N}$, are *dense*[1] in the interval $[0, 2\pi]$ and, thus, we can approximate to arbitrary precision any rotation as:

$$\mathcal{R}_n(\beta) \approx [\mathcal{R}_n(\alpha)]^\kappa . \tag{3.76}$$

Here we show that it is possible to approximate any single-qubit gate $\hat{U}$ by just using two gates, namely, the $\hat{T}$ gate, that, as we have seen, reads:

$$\hat{T} \to e^{i\pi/8} \begin{pmatrix} e^{-i\pi/8} & 0 \\ 0 & e^{i\pi/8} \end{pmatrix}, \tag{3.77}$$

and the *square root* of the Hadamard gate, namely:

$$\sqrt{\mathbf{H}} \to e^{i\pi/4} \begin{pmatrix} e^{-i\pi/4} & e^{-i\pi/4} \\ e^{-i\pi/4} & e^{i\pi/4} \end{pmatrix}. \tag{3.78}$$

Note that:

$$\sqrt{\mathbf{H}} = \frac{e^{i\pi/4}}{\sqrt{2}} (\hat{\mathbb{I}} - i\,\mathbf{H}) \quad \text{and} \quad (\sqrt{\mathbf{H}})^{-1} = \frac{e^{-i\pi/4}}{\sqrt{2}} (\hat{\mathbb{I}} + i\,\mathbf{H}) . \tag{3.79}$$

Thereafter, our problem is to find the two orthogonal axes, n and m, and an irrational angle, α, to achieve the corresponding rotations by suitable combinations of $\hat{T}$ and $\sqrt{\mathbf{H}}$.

To perform the calculations, it is useful to formally rewrite $\hat{T}$ as function of the Pauli matrices. We have:

$$T \to e^{i\pi/8} \left(\cos\frac{\pi}{8}\,\mathbb{1} - i \sin\frac{\pi}{8}\,\sigma_z \right) \tag{3.80a}$$

$$= e^{i\pi/8} \exp\left(-i\frac{\pi}{8}\,\sigma_z\right) \tag{3.80b}$$

$$= \left[e^{i\pi/2} \exp\left(-i\frac{\pi}{2}\,\sigma_z\right) \right]^{\frac{1}{4}} \to \sqrt[4]{\hat{\sigma}_z} . \tag{3.80c}$$

Therefore we also have (remember that we should use only the two chosen gates as starting point):

$$\sqrt{\mathbf{H}}\sqrt{\mathbf{H}}\,\hat{T}\,\sqrt{\mathbf{H}}\sqrt{\mathbf{H}} \equiv \sqrt[4]{\hat{\sigma}_x} . \tag{3.81}$$

---

[1] A subset of real numbers $\mathbb{A}$ is dense in the subset of real numbers $\mathbb{B}$ if $\forall \beta \in \mathbb{B}$ there exist $\alpha \in \mathbb{A}$ and $\varepsilon > 0$ such that $|\beta - \alpha| < \varepsilon$.

3.8 Universal Quantum Gates

Now we consider the unitary operation given by:

$$\hat{T}^{-1} \sqrt{\mathbf{H}}\sqrt{\mathbf{H}} \hat{T} \sqrt{\mathbf{H}}\sqrt{\mathbf{H}} = (\sqrt[4]{\hat{\sigma}_z})^{-1} \sqrt[4]{\hat{\sigma}_x} \quad (3.82a)$$

$$= \exp\left(i\frac{\pi}{8} \sigma_z\right) \exp\left(-i\frac{\pi}{8} \sigma_x\right) \quad (3.82b)$$

$$= \cos^2 \frac{\pi}{8} \mathbb{1} - i \sin \frac{\pi}{8} \, \mathbf{q} \cdot \boldsymbol{\sigma} \quad (3.82c)$$

$$= \cos \alpha \, \mathbb{1} - i \sin \alpha \, \mathbf{n} \cdot \boldsymbol{\sigma}, \quad (3.82d)$$

where

$$\mathbf{n} = \frac{\mathbf{q}}{|\mathbf{q}|} \quad \text{with} \quad \mathbf{q} = \left(\cos\frac{\pi}{8}, -\sin\frac{\pi}{8}, -\cos\frac{\pi}{8}\right), \quad (3.83)$$

and we defined:

$$\cos \alpha = \cos^2\left(\frac{\pi}{8}\right) = \frac{1}{2}\left(1 + \frac{1}{\sqrt{2}}\right), \quad (3.84)$$

α thus being an irrational multiple of 2π.

The second rotation we are looking for can be obtained as:

$$(\sqrt{\mathbf{H}})^{-1} \left(\hat{T}^{-1} \sqrt{\mathbf{H}}\sqrt{\mathbf{H}} \hat{T} \sqrt{\mathbf{H}}\sqrt{\mathbf{H}}\right) \sqrt{\mathbf{H}} \to \cos \alpha \, \mathbb{1} - i \sin \alpha \, \mathbf{m} \cdot \boldsymbol{\sigma}, \quad (3.85)$$

where $\mathbf{m} = \mathbf{p}/|\mathbf{p}|$ with:

$$\mathbf{p} = \frac{1}{\sqrt{2}}\left(\cos\left(\frac{\pi}{8} + \frac{\pi}{2}\right), -2\sin\left(\frac{\pi}{8} + \frac{\pi}{2}\right), -\cos\left(\frac{\pi}{8} + \frac{\pi}{2}\right)\right), \quad (3.86)$$

α is the same angle introduced above and we used:

$$(\sqrt{\mathbf{H}})^{-1} \hat{\sigma}_x \sqrt{\mathbf{H}} = \frac{\hat{\sigma}_x - \sqrt{2}\,\hat{\sigma}_y + \hat{\sigma}_z}{2}, \quad (3.87a)$$

$$(\sqrt{\mathbf{H}})^{-1} \hat{\sigma}_y \sqrt{\mathbf{H}} = \frac{\hat{\sigma}_x - \hat{\sigma}_z}{\sqrt{2}}, \quad (3.87b)$$

$$(\sqrt{\mathbf{H}})^{-1} \hat{\sigma}_z \sqrt{\mathbf{H}} = \frac{\hat{\sigma}_x + \sqrt{2}\,\hat{\sigma}_y + \hat{\sigma}_z}{2}. \quad (3.87c)$$

Note that $\mathbf{n} \cdot \mathbf{m} = 0$.

In summary, we have found the two rotations:

$$\mathcal{R}_n(\alpha) = \cos\alpha\, \mathbb{1} - i\sin\alpha\, \boldsymbol{n}\cdot\boldsymbol{\sigma}\,, \tag{3.88}$$

$$\mathcal{R}_m(\alpha) = \cos\alpha\, \mathbb{1} - i\sin\alpha\, \boldsymbol{m}\cdot\boldsymbol{\sigma}\,, \tag{3.89}$$

needed to approximate to arbitrary precision any single-qubit unitary (3.76).

If we consider **H** instead of $\sqrt{\textbf{H}}$, one can choose the two rotations:

$$\textbf{H}\hat{T}\textbf{H}\hat{T} = \sqrt[4]{\hat{\sigma}_x}\sqrt[4]{\hat{\sigma}_z}$$

and

$$\textbf{H}(\textbf{H}\hat{T}\textbf{H}\hat{T})\textbf{H} = \sqrt[4]{\hat{\sigma}_z}\sqrt[4]{\hat{\sigma}_x},$$

but they are not orthogonal and we cannot use Eq. (3.75), thus requiring a decomposition of the unitary operator in more than three factors.

3.9 Universality and Quantum Advantage

In the previous section we discussed the universality of a given set of gates. More in general, one of the properties of a set of quantum gates to be universal is that it should contains more than the so-called *Clifford group* {CNOT, **H**, $\hat{S}$}. We have a theorem, proved by Daniel Gottesman and Emanuel Knill, showing that a quantum computation based only on Clifford gates and measurements of observables in the Pauli group can be efficiently simulated on a classical computer:

Theorem 3.1 (Gottesman–Knill Theorem) *Any quantum computer performing only:*

(a) Clifford group gates;
(b) measurements of Pauli group operators;
(c) Clifford group operations conditioned on classical bits, which may be the results of earlier measurements;

can be perfectly simulated in polynomial time on a probabilistic classical computer.

Therefore, to have a quantum advantage, we need to go beyond the Clifford group and this still holds also in the presence of the entanglement!

As a matter of fact, not all the quantum algorithms can be implemented by using only Clifford gates, but there are ones, such as the quantum teleportation, that fall into the requests of the Gottesman–Knill theorem and, thus, can be efficiently simulated on a classical computer. We will see in Chap. 8 that also certain classes of quantum error-correcting codes can be described within this framework.

Problems

3.1 ♣ (Entanglement swapping) Consider the entangled state:

$$|\psi_{AB}\rangle = \frac{|0_A\rangle|0_B\rangle + |1_A\rangle|1_B\rangle}{\sqrt{2}},$$

belonging to the Hilbert space $\mathcal{H}_A \otimes \mathcal{H}_B$ and show that if we teleport the state of the qubit A onto the qubit D of the entangled state:

$$|\psi_{CD}\rangle = \frac{|0_C\rangle|0_D\rangle + |1_C\rangle|1_D\rangle}{\sqrt{2}},$$

belonging to the Hilbert space $\mathcal{H}_C \otimes \mathcal{H}_D$, then we get the entangled state:

$$|\psi_{AD}\rangle = \frac{|0_A\rangle|0_D\rangle + |1_A\rangle|1_D\rangle}{\sqrt{2}}.$$

This protocol is known as "entanglement swapping", since the entanglement between the systems A–B and C–D is eventually swapped between the two previously uncorrelated systems.

3.2 ♣ (Deutsch problem) Find four possible two-qubit unitary operators $\hat{U}_f$ implementing the corresponding $f : \{0, 1\} \to \{0, 1\}$, namely:

(a) $f(0) = f(1) = 0$;

(b) $f(0) = f(1) = 1$;

(c) $f(0) \neq f(1) = 0$;

(d) $f(0) \neq f(1) = 1$.

3.3 ♣ Prove that the function:

$$f(x) = a \cdot x \equiv a_0 x_0 \oplus \cdots \oplus a_{n-1} x_{n-1}$$

of the Bernstein–Vazirani problem can be only constant or balanced.

Further Readings

D. Gottesman, *The Heisenberg Representation of Quantum Computers*, e-Print: quant-ph/9807006 [quant-ph]

C.-K. Li, R. Roberts, X. Yin, Decomposition of unitary matrices and quantum gates. Int. J. Quantum Inf. **11**, 1350015 (2013)

N.D. Mermin, *Quantum Computer Science* (Cambridge University Press, 2007) – Chapter 2

M.A. Nielsen, I.L. Chuang, *Quantum Computation and Quantum Information* (Cambridge University Press, 2010) – Chapter 1, Chapter 4.5, Chapter 10.6

J. Stolze, D. Suter, *Quantum Computing: A Short Course from Theory to Experiment* (Wiley-VCH, 2008) – Chapter 5.3

Universal Computers and Computational Complexity

4

Abstract

In this chapter we (really) briefly describe two important examples of "universal computers": the *Turing machine* and its quantum counterpart, the *quantum Turing machine*. These "machines" are useful to check computability and efficiency of algorithms without specifying a particular hardware implementation, that is one of the main tasks of computer science. For the sake of completeness we also introduce the main complexity classes (P, NP and their quantum analogue BQP and QMA).

4.1 The Turing Machine

The Turing machine is the simplest example for a *universal quantum computer*. We define the deterministic Turing machine as a discrete-time dynamical system, with an infinite input/output tape, a head to read and write symbols on the tape and a set of internal state. To these states belongs also the so-called "halt" state, in which the machine stops.

The configuration of a Turing machine can be defined as $C = (q_C, h_C, s_C)$, where q_C is the internal state of the machine, h_C the head position and s_C is the string of the symbols on the tape. The particular symbol at position h on the tape, is given by $s(h)$ (see Fig. 4.1). The classical transition rules, given an internal state p and a read symbol σ, can be defined as $\delta_c(p, \sigma) = (\tau, q, d)$, namely, the machine writes the symbol τ on the tape, changes its internal state to q and moves the head position from h to $h + d$, where $d = +1$ ($d = -1$) corresponds to a step to the right (left). If, for instance, $\delta_c(q_C, s_C(h_C)) = (\tau, q, d)$, we have the transition:

$$(q_C, h_C, s_C) \to (q, h_C + d, s_C^\tau), \tag{4.1}$$

© The Author(s), under exclusive license to Springer Nature Switzerland AG 2025
S. Olivares, *A Student's Guide to Quantum Computing*, Lecture Notes in Physics 1038, https://doi.org/10.1007/978-3-031-83361-8_4

Fig. 4.1 Schematic view of a Turing machine

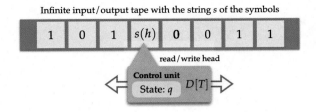

where $s_C^\tau(h_C) = \tau$ whereas $s_C^\tau(j) = s_C(j)$ for $j \neq h_C$. A computation is carried out by a suitable definition of the transition rules and the number of steps required to complete the computation is related to the complexity of the problem.

Though the Turing machine has no a practical interest, it is worth noting that every task that can be performed by a computer can be performed by a Turing machine. In fact, according to the Church–Turing thesis, *every function which would be naturally regarded as computable can be computed by the universal Turing machine* (there is also a strong Church–Turing thesis: *Any model of computation can he simulated on a probabilistic Turing machine with at most a polynomial increase in the number of elementary operations required*). A universal Turing machine T_U can simulate every Turing machine T given its description $D[T]$, namely, a binary number encoding the set of its transition rules. In particular, the number of steps used by T_U to simulate a given T is a polynomial function of the length of the number $D[T]$. If x is the input and $T(x)$ the output of the Turing machine T, then $T_U(D[T], x) = T(x)$.

4.2 The Quantum Turing Machine

The quantum version of the Turing machine, the quantum Turing machine, is characterized by a quantum state $|C\rangle$ corresponding to its configuration, that is a vector in a suitable Hilbert space. We can write the configuration as:

$$|C\rangle = |q_C\rangle|h_C\rangle|s_C\rangle. \tag{4.2}$$

The reader can see that the internal state, the head position and the symbol on the tape are substituted by quantum states. The evolution, now, is not deterministic, but is described by a unitary U determined by the quantum transition function:

$$\delta(p, \sigma; \tau, q, d) \in \mathbb{C}, \tag{4.3}$$

that is the amplitude of the classical transition $(p, \sigma) \to (\tau, q, d)$. Therefore, we have:

$$U|C\rangle = \sum_{\tau, q, d} \delta(q_C, s_C(h_C); \tau, q, d) |q\rangle|h_C + d\rangle|s_C^\tau\rangle. \tag{4.4}$$

The result of the computation is obtained by measuring the tape state after the computation has been completed. And this is an issue, since the computation is completed when the internal state $|q\rangle$ is found into the halt state, but to check it one should perform a measurement which may disturb (and usually it does) the computation itself. In order to solve the problem, Deutsch proposed to introduce an additional qubit, the *halt qubit*, and an observable $\hat{n}_0$ to monitor it. When the internal state is different from the halt state, the halt qubit is $|0\rangle$, but in the presence of the halt state, its value is $|1\rangle$. Therefore, one initializes the halt qubit to $|0\rangle$ and a valid algorithm sets its value to $|1\rangle$ only at the end of the computation, without interacting with it otherwise. The observable $\hat{n}_0$, according to Deutsch, can be periodically observed from the outside, without affecting the operation of the machine.

4.3 Important Classical and Quantum Complexity Classes

In this section we focus on the so-called *decision problems*, namely, problems whose question can be posed as a yes-no question. Nevertheless, decision problems are closely related to *function problems*, where the problem is to compute the values of a given function. The space containing decision problems that can be solved by a Turing machine using a polynomial amount of space is called PSPACE (if n is the number of bits of the input, $O(n^k)$ is the amount of needed memory).

In general, a computational problem can be classified according to several measures of complexity. The thorough analysis of complexity theory is beyond the scope of these notes and here we mention only some important classes (the interested reader can find a thorough analysis of the complexity classes zoo in the Further Readings at the end of this chapter.).

Let us consider a task to be performed on an integer number x (the input of our computation). Depending on the particular task, a Turing machine will (hopefully) require a given number of steps s to solve it. As a matter of fact, s depends on x (factorizing a small integer requires less steps than the factorization of a 20-digit integer!). The computational complexity of the task characterizes how the number s increases with the number of bit needed to encode x, namely, $L = \log_2 x$. For instance, if the task is the calculation of x^2, we can find *an algorithm* such that, roughly, $s \sim L^2$. When s is a *polynomial* function of L, as in the latter case, we say that the problem belongs to the complexity class P (*polynomial in time*). If s rises exponentially with L, the problem is considered *hard*.

However, in many cases verify the solution is much easier than to find it. It is the case of the factorization of large integers: for the best algorithm we currently know we have:

$$s \sim \exp\left(\sqrt[3]{\frac{64}{9} L \log L}\right). \tag{4.5}$$

This kind of decision problems belongs to the class NP, where NP means *nondeterministic polynomial*. Furthermore, the problem is NP-hard if

$$s \sim e^{f(L)}, \tag{4.6}$$

fore some $f(L) > 0$.

A nondeterministic polynomial algorithm may at any step follows two different paths which are followed in parallel: one can perform an exponential number of calculations in polynomial time (at the expense of the computational capacity, that grows exponentially...). In order to verify the solution one simply follows, in polynomial time, the right path of the tree-like algorithm.

We can give a more formal definition of the NP class as follows. A decision problem P belongs to the NP class if, given the input string (of bits) x, an *omniscient prover* can provide a proof π (another string of bits) of P such that a verifier can run a deterministic algorithm V in polynomial time to verify it, namely, $V(x, \pi) = 1$ if x is a yes-instance or $V(x, \pi) = 0$ if x is a no-instance. We should also meet two further properties. The first one is called *soundness*: if x is a yes-instance, then there must exist a prove π such that $V(x, \pi) = 1$; the other properties is the *completeness*, that is if x is a no-instance then $\forall \pi$ one finds $V(x, \pi) = 0$.

There is a class of problems, called Merlin–Arthur, MA, that is connected to the NP one. In this case, given the instance x, the prover, Merlin, gives a proof π to the verifier, Arthur, that can run a *probabilistic* poly-time algorithm V_p. Now, if x is a yes-instance, then there exists a proof π such that $V_p(x, \pi) = 1$ with a probability $p \geq 2/3$ (or, sometimes, $p \geq 3/4$); if, on the other hand, x is a no-instance, $\forall \pi$ provided by Merlin we find $V_p(x, \pi) = 1$ with probability $p < 1/3$ or (or, sometimes, $p < 1/4$).

Another important complexity class contains the so-called NP-complete problems. We recall that a problem P_2 is *reducible* or *polynomially reducible* to a problem P_1, if the solution of P_2 can be found by applying P_1 a polynomial number of times (one also say that "P_2 cannot be harder than P_1"). A problem is NP-complete if any NP problem can be reduced to it. A typical NP-complete problem is the *travelling salesman problem*: given a list of cities and the distances between each pair of them, find the shortest possible route that visits each city exactly once and returns to the origin city. More precisely, the class NP-complete corresponds to the intersection between the class NP and the class NP-hard, the latter being the set of problems (not only decision problems, but also optimization problems and so on). We can also say that a NP-hard problem is at least as hard as the hardest problems in NP. One of the most fundamental problems of theoretical quantum computer science is to find whether P and NP coincide: if somebody finds a polynomial solution for any NP-complete problem, then P = NP.

While the *travelling salesman problem* is an optimization problem belonging to both NP-hard and NP-complete, there are decision problems that are NP-hard but not NP-complete, for example the *halting problem*: suppose to have a Turing machine T with description $D[T]$, will the machine stop for a given input x? It is possible to show that NP-complete problems can be reduced to this problem; but it is also

4.3 Important Classical and Quantum Complexity Classes

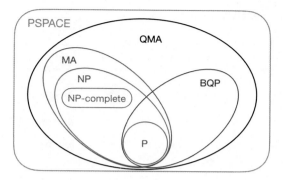

Fig. 4.2 Relation between some important complexity classes discussed in the text

well-known that it is an example of *unsolvable* problem, thus, not complete! This can be proved by supposing that there exists a universal Turing machine T_H with description $D[T_H]$ such that "$T_H(D[T])$ halts iff $T(D[T])$ does not halt" (we are using as input for the machines binary number $D[T]$). But if we put now $T \equiv T_H$ we have the clear contradiction "$T_H(D[T_H])$ halts iff $T_H(D[T_H])$ does not halt"!! This is the argument used by the same Turing to prove that a general algorithm to decide whether a Turing machine stops does not exist.

In Fig. 4.2 we show a pictorial view of the relation between the complexity classes P, NP and NP-complete. In the same figure we also report the two most important quantum complexity classes: the class BQP (*bound-error quantum polynomial time*) and the class QMA (*quantum Merlin Arthur*). The BQP is the class of decision problems which can be solved by a quantum computer in polynomial time, with an error probability of at most 1/3 for all instances. To this class belongs, in particular, the Shor's factorization algorithm: it requires

$$s \sim L^2 \log L \, \log \log L, \tag{4.7}$$

this demonstrating that the integer factorization problem can be efficiently solved on a quantum computer and, thus, belongs to the complexity class BQP.

The last class we mention is the QMA, the quantum analog of the class NP: it is related to BQP in the same way NP is related to P. Roughly speaking, the class QMA contains the decision problems for which the proofs, given by the oracle, Merlin, with infinite power, should be verifiable by Arthur in polynomial time on a quantum computer through a suitable algorithm V_q. As in the case of the MA class, if x is a yes-instance, then there exists a proof associated with the state $|\varphi\rangle$, such that $V_q(x, |\varphi\rangle) = 1$ with probability $p \geq 2/3$; if x is a no-instance, then $\forall |\varphi\rangle$ provided by the oracle, we find $V_q(x, |\varphi\rangle) = 1$ with probability $p \leq 1/3$.

Further Readings

D. Deutsch, Quantum theory, the Church-Turing principle and the universal quantum computer. Proc. R. Soc. London A **400**, 97–117 (1985)

M.A. Nielsen, I.L. Chuang, *Quantum Computation and Quantum Information* (Cambridge University Press, 2010) – Chapter 3.2, Chapter 4.5.5

M. Ozawa, Quantum nondemolition monitoring of universal quantum computers. Phys. Rev. Lett. **80**, 631–634 (1998)

J. Stolze, D. Suter, *Quantum Computing: A Short Course from Theory to Experiment* (Wiley-VCH, 2004) – Chapter 3.3

J. Watrous, in *Encyclopedia of Complexity and Systems Science*, ed. by R. Meyers (Springer, New York, 2009)

Quantum Fourier Transform and Shor's Factoring Algorithm

5

Abstract

In this chapter we introduce the Quantum Fourier Transform (QFT), which is a key ingredient of many quantum protocols, and estimate the number of quantum gates to implement it. Then, we apply the QFT to the phase estimation problem and address the factoring algorithm proposed by Shor. In particular, we highlight the role of the QFT in the order-finding protocol that allows overcoming the computational limits of the best classical algorithms.

5.1 Discrete Fourier Transform and QFT

Though its meaning is different form the quantum counterpart, it is useful to describe the action of the (classical) discrete Fourier transform. As the reader will see, there are some analogies between the classical and quantum versions, but an deeper insight clearly reveals their fundamental differences.

The (classical) discrete Fourier transform maps a vector $(x_1, \ldots, x_N)$ of N complex numbers into a new vector $(y_1, \ldots, y_N)$, where:

$$y_h = \frac{1}{\sqrt{N}} \sum_{k=1}^{N} \exp\left(2\pi i \frac{h\,k}{N}\right) x_k. \tag{5.1}$$

In a "similar" way we can define the QFT. Given the n-qubit state:

$$|x\rangle_n = \bigotimes_{m=0}^{n-1} |x_m\rangle \tag{5.2}$$

$$= |x_{n-1}\rangle|x_{n-2}\rangle \ldots |x_0\rangle, \tag{5.3}$$

© The Author(s), under exclusive license to Springer Nature Switzerland AG 2025
S. Olivares, *A Student's Guide to Quantum Computing*, Lecture Notes in Physics 1038, https://doi.org/10.1007/978-3-031-83361-8_5

where x is an integer number, $0 \le x < 2^n$, and $x_{n-1}x_{n-2}\ldots x_0$ is its binary representation, namely, $x = \sum_{k=0}^{n-1} x_k 2^k$, with $x_k \in \{0, 1\}$, we have:

$$\hat{F}_Q |x\rangle_n = \frac{1}{2^{n/2}} \sum_{y=0}^{2^n-1} \exp\left(2\pi i \frac{x\,y}{2^n}\right) |y\rangle_n. \tag{5.4}$$

Since $|y\rangle_n = \bigotimes_{m=0}^{n-1} |y_m\rangle$ and $y = \sum_{m=0}^{n-1} y_m 2^m$, we can write Eq. (5.4) as:

$$\hat{F}_Q |x\rangle_n = \frac{1}{2^{n/2}} \sum_{y_{n-1}=0}^{1} \cdots \sum_{y_0=0}^{1} \bigotimes_{m=0}^{n-1} \exp\left(2\pi i \frac{x\,y_m}{2^{n-m}}\right) |y_m\rangle, \tag{5.5}$$

$$= \frac{1}{2^{n/2}} \bigotimes_{m=0}^{n-1} \left[|0_m\rangle + \exp\left(2\pi i \frac{x}{2^{n-m}}\right) |1_m\rangle \right], \tag{5.6}$$

$$= \bigotimes_{m=0}^{n-1} |\psi_m\rangle, \tag{5.7}$$

where we defined:

$$|\psi_m\rangle = \frac{1}{\sqrt{2}} \left[|0_m\rangle + \exp\left(2\pi i \frac{x}{2^{n-m}}\right) |1_m\rangle \right].$$

In Fig. 5.1 we show the action of the QFT on the state $|x\rangle_n$.

In order to find the quantum circuit implementing the QFT, instead of the transformation (5.7) it is better to consider the following one (for the sake of simplicity we use the same symbol $\hat{F}_Q$ for both the operations):

$$\hat{F}_Q |x\rangle_n = \frac{1}{2^{n/2}} \bigotimes_{m=1}^{n} \left[|0_{n-m}\rangle + \exp\left(2\pi i \frac{x}{2^m}\right) |1_{n-m}\rangle \right], \tag{5.8a}$$

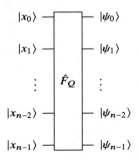

Fig. 5.1 Quantum Fourier transform: the input n-qubit state $|x\rangle_n = \bigotimes_{k=0}^{n-1} |x_k\rangle = |x_{n-1}\rangle \ldots |x_0\rangle$ is transformed into the output n-qubit state $\bigotimes_{k=0}^{n-1} |\psi_k\rangle = |\psi_{n-1}\rangle \ldots |\psi_0\rangle$. See the text for details

5.1 Discrete Fourier Transform and QFT

$$= \frac{1}{2^{n/2}} \bigotimes_{m=0}^{n-1} \left[|0_{n-m-1}\rangle + \exp\left(2\pi i \frac{x}{2^{m+1}}\right) |1_{n-m-1}\rangle \right], \quad (5.8b)$$

$$= \bigotimes_{m=0}^{n-1} |\psi_{n-m-1}\rangle. \quad (5.8c)$$

The subtle difference between (5.7) and (5.8a) is that the overall action of the first one can be summarized as:

$$|x_0\rangle \rightarrow |\psi_0\rangle,$$
$$|x_1\rangle \rightarrow |\psi_1\rangle,$$
$$\vdots$$
$$|x_{n-1}\rangle \rightarrow |\psi_{n-1}\rangle,$$

while in the second case we have:

$$|x_0\rangle \rightarrow |\psi_{n-1}\rangle,$$
$$|x_1\rangle \rightarrow |\psi_{n-2}\rangle,$$
$$\vdots$$
$$|x_{n-1}\rangle \rightarrow |\psi_0\rangle,$$

or, in a more compact form (for the sake of simplicity we drop the subscripts):

$$|x_m\rangle \rightarrow |\psi_{n-m-1}\rangle = \frac{1}{\sqrt{2}} \left[|0\rangle + \exp\left(2\pi i \frac{x}{2^{m+1}}\right) |1\rangle \right]. \quad (5.9)$$

Note that we can also write:

$$\exp\left(2\pi i \frac{x}{2^{m+1}}\right) = \prod_{k=0}^{n-1} \exp\left(2\pi i \frac{x_k 2^k}{2^{m+1}}\right), \quad (5.10)$$

where we used $x = \sum_{k=0}^{n-1} x_k 2^k$. By introducing the function:

$$f_m(z, k) = \begin{cases} \exp\left(2\pi i \frac{z}{2^{m-k+1}}\right) & \text{if } 0 \le k < m, \\ (-1)^z & \text{if } k = m, \\ 1 & \text{if } m < k < n, \end{cases} \quad (5.11)$$

with $z \in \{0, 1\}$, we have:

$$|x_m\rangle \rightarrow \frac{1}{\sqrt{2}} \left[|0\rangle + \prod_{k=0}^{n-1} f_m(x_k, k) |1\rangle \right] \qquad (5.12)$$

$$= \frac{1}{\sqrt{2}} \left[|0\rangle + \prod_{k=0}^{m} f_m(x_k, k) |1\rangle \right]. \qquad (5.13)$$

If we now define the operator $\hat{R}_h(z)$, such that:

$$\hat{R}_h(z)|0\rangle = |0\rangle, \quad \text{and} \quad \hat{R}_h(z)|1\rangle = \exp\left(2\pi i \frac{z}{2^h}\right) |1\rangle, \qquad (5.14)$$

which corresponds to the 2×2 matrix:

$$\hat{R}_h(z) \rightarrow \begin{pmatrix} 1 & 0 \\ 0 & \exp\left(2\pi i \frac{z}{2^h}\right) \end{pmatrix}, \qquad (5.15)$$

we can write (for $m > 0$):

$$|x_m\rangle \rightarrow \frac{1}{\sqrt{2}} \left[|0\rangle + \prod_{k=0}^{m} f_m(x_k, k) |1\rangle \right] \qquad (5.16)$$

$$= \hat{R}_{m+1}(x_0) \hat{R}_m(x_1) \ldots \hat{R}_2(x_{m-1}) \underbrace{\frac{|0\rangle + (-1)^{x_m}|1\rangle}{\sqrt{2}}}_{\mathbf{H}|x_m\rangle}, \qquad (5.17)$$

where **H** is the Hadamard transformation (see Sect. 1.4.4). In order to be clearer, we can look at the evolution of the first three qubits:

$$|x_0\rangle \rightarrow \frac{|0\rangle + f_1(x_0, 0)|1\rangle}{\sqrt{2}} \equiv \mathbf{H}|x_0\rangle, \qquad (5.18)$$

$$|x_1\rangle \rightarrow \frac{|0\rangle + f_1(x_0, 0) f_1(x_1, 1)|1\rangle}{\sqrt{2}} \qquad (5.19)$$

$$= \frac{|0\rangle + \hat{R}_2(x_0)(-1)^{x_1}|1\rangle}{\sqrt{2}} \equiv \hat{R}_2(x_0)\mathbf{H}|x_1\rangle, \qquad (5.20)$$

$$|x_2\rangle \rightarrow \frac{|0\rangle + f_2(x_0, 0) f_2(x_1, 1) f_2(x_2, 2)|1\rangle}{\sqrt{2}} \qquad (5.21)$$

5.1 Discrete Fourier Transform and QFT

$$= \frac{|0\rangle + \hat{R}_3(x_0)\hat{R}_2(x_1)(-1)^{x_2}|1\rangle}{\sqrt{2}} \equiv \hat{R}_3(x_0)\hat{R}_2(x_1)\mathbf{H}|x_2\rangle, \quad (5.22)$$

where we used Eq. (5.11). More in general, if $0 < m < n$ we have:

$$|x_m\rangle \rightarrow \prod_{k=0}^{m-1} \hat{R}_{m-k+1}(x_k)\, \mathbf{H}|x_m\rangle.$$

As a matter of fact, $\hat{R}_h(0) = \hat{\mathbb{I}}$, thus we can see $\hat{R}_h(x_k)$ as a *controlled* gate, which applies a phase shift to the corresponding qubit only if the control qubit $|x_k\rangle$ assumes the value $x_k = 1$. Therefore, the corresponding quantum circuit involves single-qubit gates (Hadamard transformations) and two-qubit gates [controlled $\hat{R}_h \equiv \hat{R}_h(1)$], as depicted in Fig. 5.2.

In order to reverse the order of the outputs, one should apply at most $n/2$ SWAP gates (recall that three CNOT gates are needed to implement a single SWAP). Besides the SWAPs, the total number of gates involved in Fig. 5.2 is:

$$n + (n-1) + \cdots + 1 = n(n+1)/2 \sim n^2. \quad (5.23)$$

Note that the classical Fast Fourier Transform algorithm needs $\sim n\, 2^n$ gates (since it ignores trivial operations such as the multiplication by 1), while the direct calculation of the discrete Fourier transform requires $\sim 2^{2n}$ gates! However, there are two main issues we should point out: (1) the final amplitudes cannot be accessed directly; (2) there is not an efficient preparation of the initial state. Finding applications of the QFT is more subtle than one might hope...

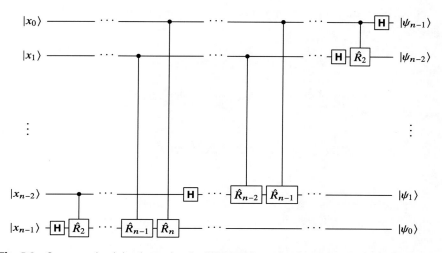

Fig. 5.2 Quantum circuit implementing the QFT (we do not implement the final SWAP gates)

5.2 The Phase Estimation Protocol

The *phase estimation* procedure is a key ingredient for many quantum algorithms. Suppose that $\hat{U}$ is an unitary operator and $|u\rangle$ is one of its eigenvectors, such that:

$$\hat{U}|u\rangle = \exp(2\pi i \phi)|u\rangle, \qquad (5.24)$$

where $\phi \in [0, 1)$ is unknown. The binary representation of ϕ is given by $0.\varphi_1\varphi_2\varphi_3\ldots$, where $\varphi_k \in \{0, 1\}$, and:

$$\phi = \sum_k \varphi_k 2^{-k}. \qquad (5.25)$$

Since ϕ is an overall phase, we cannot directly retrieve it. However, if we have "black boxes" (the *oracles*) capable of preparing $|u\rangle$ and of performing the controlled-$\hat{U}^{2^{k-1}}$ operations, namely $c\hat{U}_k^{2^{k-1}}$, which use the k-th qubit as control, we can succeed in the estimation of ϕ. Note that since we cannot access $\hat{U}$ (for this reason it is represented as a "black box"), the phase estimation procedure is not a complete algorithm in its own right.

At first, we assume that ϕ can be exactly specified with n bits: in this case the estimation procedure allows us to obtain the actual value ϕ. The protocol uses two registers: the first one contains n qubits prepared in the initial state $|0\rangle_n$; the second one contains many qubit as is necessary to store $|u\rangle$ (without loss of generality we assume that only one qubit is needed). The first step of the procedure applies n Hadamard transformations to $|0\rangle_n$, generating a balanced superposition of all the states $|x\rangle_n$, $0 \le x < 2^n$. Then we apply controlled-$\hat{U}^{2^{k-1}}$ to $|u\rangle$ with control qubit corresponding to the k-th qubit of the first register (see Fig. 5.3).

Since the action of the controlled $c\hat{U}_k^{2^{n-k}}$ is (note, also in this case, the use of the phase kickback):

$$c\hat{U}_k^{2^{n-k}}|x_k\rangle|u\rangle = \exp\left(2\pi i \phi\, x_k\, 2^{n-k}\right)|x_k\rangle|u\rangle, \qquad (5.26)$$

we have (we write only the evolution of the k-th qubit of the first register and the second register):

$$c\hat{U}_k^{2^{k-1}}(\mathbf{H} \otimes \hat{\mathbb{I}})|0_k\rangle|u\rangle = \frac{1}{\sqrt{2}}\left[|0_k\rangle + \exp\left(2\pi i\, 2^{k-1}\phi\right)|1_k\rangle\right]|u\rangle \equiv |\psi_k\rangle|u\rangle. \qquad (5.27)$$

Therefore, after the first step of the procedure, the first register evolves as follows (since the second register is left unchanged, we do not write it explicitly):

$$|0\rangle_n \to |\psi_n\rangle|\psi_{n-1}\rangle\ldots|\psi_1\rangle \equiv |\Psi(\phi)\rangle_n. \qquad (5.28)$$

5.2 The Phase Estimation Protocol

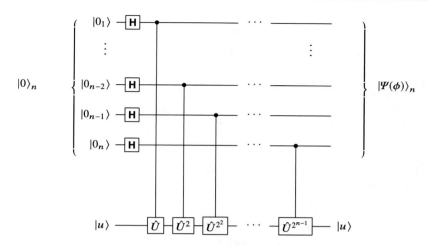

Fig. 5.3 Quantum circuit representing the first step of the phase estimation procedure. The expression of the state $|\Psi(\phi)\rangle_n$ is given in Eq. (5.29)

As in the case of Eq. (5.7), we can write:

$$|\Psi(\phi)\rangle_n = \frac{1}{2^{n/2}} \sum_{x=0}^{2^n-1} \exp(2\pi i \phi x) |x\rangle_n, \tag{5.29}$$

where it is worth noting that here $x = \sum_{k=1}^{n} x_k 2^{k-1}$. Now we apply the inverse of the QFT to $|\Psi(\phi)\rangle_n$:

$$\hat{F}_Q^\dagger |\Psi(\phi)\rangle_n = \frac{1}{2^n} \sum_{x=0}^{2^n-1} \exp(2\pi i \phi x) \sum_{y=0}^{2^n-1} \exp\left(-2\pi i \frac{yx}{2^n}\right) |y\rangle_n \tag{5.30a}$$

$$= \frac{1}{2^n} \sum_{y=0}^{2^n-1} \sum_{x=0}^{2^n-1} \exp\left[-2\pi i x \frac{(y-2^n\phi)}{2^n}\right] |y\rangle_n \tag{5.30b}$$

$$\underbrace{\phantom{\frac{1}{2^n} \sum_{y=0}^{2^n-1} \sum_{x=0}^{2^n-1} \exp\left[-2\pi i x \frac{(y-2^n\phi)}{2^n}\right]}}_{2^n \delta_{0,y-2^n\phi}}$$

$$= |2^n \phi\rangle_n \equiv |\varphi\rangle_n \tag{5.30c}$$

where in Eq. (5.30c) we defined the *integer* number φ as:

$$2^n \phi = 2^n \sum_{m=1}^{n} \varphi_m 2^{-m} = \sum_{k=0}^{n-1} \varphi_{n-k} 2^k \equiv \varphi, \tag{5.31}$$

Fig. 5.4 Phase estimation with the T or $\frac{\pi}{8}$ gate. To obtain the integer number associated with the binary expansion $\varphi_3\varphi_2\varphi_1$ one should multiply φ_k by $3-k$. See the text for details

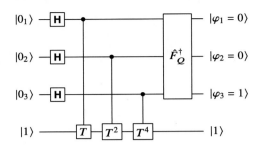

and we recall that both y and φ are integers less then 2^n [otherwise we don't have the Kronecker delta, see Eq. (5.35) below]. Finally, since (note that to obtain the corresponding integer number at the output one should multiply φ_k by $n-k$ and not k as usual):

$$|\varphi\rangle_n = |\varphi_n\rangle|\varphi_{n-1}\rangle\ldots|\varphi_1\rangle, \tag{5.32}$$

we can retrieve the value of each bit φ_k by measuring the corresponding qubit in the computational basis and obtain $\varphi = 0.\varphi_1\varphi_2\ldots\varphi_n$.

It the following example, we consider the T gate defined in Eq. (3.5). It is straightforward to verify that $T|1\rangle = e^{2\pi i \phi}|1\rangle$ with $\phi = 1/8$ or, in binary notation, $\phi = 0.\varphi_1\varphi_2\varphi_3 = 0.001_2$ (where the subscript 2 refers to the chosen basis). The quantum circuit to implement the phase estimation is drawn in Fig. 5.4. The state of the input register after the inverse of the QFT reads (the proof is left to the reader):

$$\hat{F}_Q^\dagger |\Psi(\phi)\rangle_3 = \frac{1}{2^3}\sum_{y=0}^{7}\underbrace{\sum_{x=0}^{7}\exp\left(-2\pi i x \frac{y-2^3\phi}{2^3}\right)}_{2^3\delta_{0,y-2^3\phi}}|y\rangle_3. \tag{5.33}$$

Since $2^3\phi = 1 = \varphi_1\varphi_2\varphi_3 = 001_2$, we obtain (note the reverse order!):

$$\hat{F}_Q^\dagger|\Psi\rangle = \left|2^3\phi\right\rangle_3 = |\varphi_3\rangle|\varphi_2\rangle|\varphi_1\rangle = |1_3\rangle|0_2\rangle|0_1\rangle. \tag{5.34}$$

If the actual value of the phase, say ϕ^*, cannot be exactly written with an n-bit expression, then the estimation does not give its actual value, but just an approximation. In fact, in this case $2^n\phi^*$ is not an integer and Eq. (5.30b) becomes:

$$\hat{F}_Q^\dagger|\Psi(\phi)\rangle_n = \sum_{y=0}^{2^n-1}\frac{1}{2^n}\frac{1-\exp\left[-2\pi i(y-2^n\phi^*)\right]}{1-\exp\left[-2\pi i(y-2^n\phi^*)2^{-n}\right]}|y\rangle_n,$$

$$= \sum_{y=0}^{2^n-1}f_y(\phi^*;n)|y\rangle_n, \tag{5.35}$$

5.2 The Phase Estimation Protocol

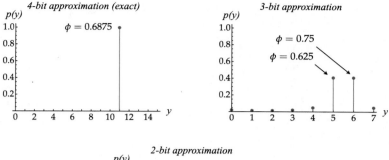

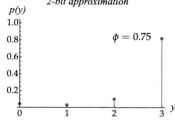

Fig. 5.5 Plot of $p(y)$ given in Eq. (5.36) for the estimation of the phase $\phi^* = 0.6875$, that has the exact binary expansion 0.1011_2. We used a different number n of qubits for the input register, from left to right: $n = 4, 3$ and 2

that is a superposition of *all* the possible outcomes $|y\rangle_n$, each with probability:

$$p(y) = |f_y(\phi^*; n)|^2 = \frac{1}{2^{2n}} \frac{1 - \cos\left[2\pi\left(y - 2^n \phi^*\right)\right]}{1 - \cos\left[2\pi\left(y - 2^n \phi^*\right)2^{-n}\right]}. \tag{5.36}$$

The reader can check that $p(y) \geq 0$ and $\sum_{y=0}^{2^n-1} p(y) = 1$. In Figs. 5.5 and 5.6 we plot the outcome probability $p(y)$ for two values of the unknown phase and a different number n of qubits of the input register.

Among the possible outcomes of the measurement there will be a particular integer $\varphi^{(b)}$, $0 \leq \varphi^{(b)} < 2^n$, such that $\phi^{(b)} = 2^{-n}\varphi^{(b)}$, is the best n-bit approximation of the actual value ϕ^*. Let us suppose that a measurement leads to the outcome φ corresponding to the phase $\phi = 2^{-n}\varphi$. One of the interesting features of the present phase-estimation procedure is that the probability that $|\varphi - \varphi^{(b)}| > t$, where the integer t represents the tolerance to error, decreases as t increases. Note that:

$$\left|\varphi - \varphi^{(b)}\right| > t \Rightarrow \left|\phi - \phi^{(b)}\right| > 2^{-n}t. \tag{5.37}$$

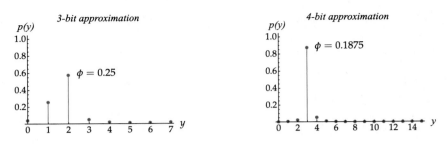
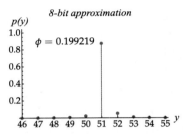

Fig. 5.6 Plot of $p(y)$ given in Eq. (5.36) for the estimation of the phase $\phi^* = 0.2$, that does not have an exact binary expansion $(0.00110011\ldots_2)$, using an increasing number n of qubits for the input register, from left to right $n = 3, 4$ and 8

It is possible to show that this probability is given by:

$$p\left(\left|\varphi - \varphi^{(b)}\right| > t\right) \leq \frac{1}{2(t-1)} \tag{5.38}$$

and, thus, the *success probability* (the probability of getting an estimation of ϕ within the tolerance t) reads:

$$p\left(\left|\varphi - \varphi^{(b)}\right| \leq t\right) > 1 - \frac{1}{2(t-1)}. \tag{5.39}$$

This result allows to calculate the number of qubits n in order to achieve the phase estimation within a given accuracy. For instance, suppose we want to approximate ϕ to an accuracy 2^{-q}, $0 < q < n$, namely:

$$\left|\phi - \phi^{(b)}\right| < 2^{-q}, \tag{5.40}$$

or, equivalently, multiplying both sides by 2^n:

$$\left|\varphi - \varphi^{(b)}\right| \leq t = 2^{n-q} - 1, \tag{5.41}$$

(note that $2^{n-q} - 1$ corresponds to the maximum integer which can be encoded using only $n - q$ bits). If we require a success probability:

$$P\left(\left|\varphi - \varphi^{(b)}\right| \leq t\right) = 1 - \varepsilon, \qquad (5.42)$$

for a given $\varepsilon > 0$, then the number n of required qubits for the first register should be at least:

$$n = q + \left\lceil \log_2\left(2 + \frac{1}{2\varepsilon}\right) \right\rceil, \qquad (5.43)$$

where $\lceil z \rceil$ is the ceiling function, which represents the smallest integer not less than $z \in \mathbb{R}$.

5.3 The Factoring Algorithm (Shor's Algorithm)

The aim of a factoring algorithm is to find the nontrivial factors of an integer N. In this section we show that the factoring problem turns out to be equivalent to the so-called *order-finding problem* we just studied, in the sense that a fast algorithm for order finding can easily be turned into a fast algorithm for factoring. The algorithm is essentially based on two theorems and it is useful to recall the following concepts. Given three integer numbers a, b and N, we have that:

$$a = b \pmod{N} \Rightarrow \exists q \in \mathbb{Z} \text{ such that } a - b = qN. \qquad (5.44)$$

Suppose, now, to have two integers, x and N, $x < N$, with *no common* factors. The *multiplicative order* or, simply, the *order* of $x \pmod{N}$ is defined to be the least positive integer r such that

$$x^r \pmod{N} = 1. \qquad (5.45)$$

For instance, given $x = 5$ and $N = 21$, we have:

$$5 \pmod{21} = 5, \quad 5^4 \pmod{21} = 16,$$
$$5^2 \pmod{21} = 4, \quad 5^5 \pmod{21} = 17,$$
$$5^3 \pmod{21} = 20, \quad 5^6 \pmod{21} = 1.$$

Therefore the order of $5 \pmod{21}$ is $r = 6$. If we consider $x = 3$ and $N = 10$, we have:

$$3 \pmod{10} = 3, \quad 3^3 \pmod{10} = 7,$$
$$3^2 \pmod{10} = 9, \quad 3^4 \pmod{10} = 1,$$

and the order of 3(mod 10) is $r = 4$. Note also that if r is the order of x modulo N, then $x^{(r+s)} \pmod{N} = x^s \pmod{N}$.

Moreover, given the Euler's totient function:

$$\varphi(N) = N \prod_{p|N} \left(1 - \frac{1}{p}\right), \tag{5.46}$$

where $p|N$ indicates the distinct prime numbers, p, dividing N, we have the Euler's theorem, that is:

$$a^{\varphi(N)} = 1 \pmod{N}, \tag{5.47}$$

with a and N co-prime.[1] Since, clearly, $\varphi(N) \leq N$, if $a^r = 1 \pmod{N}$, we have:

$$r \leq \varphi(N) \leq N. \tag{5.48}$$

We can now state the two theorems that are at the basis of the factoring algorithm.

Theorem 5.1 *Suppose N is an L-bit composite number, and x is a non-trivial solution to the equation $x^2 = 1 \pmod{N}$ in the range $1 \leq x \leq N$, that is, $x \neq \pm 1 \pmod{N}$. Then at least one of $\gcd(x-1, N)$ and $\gcd(x+1, N)$ is a non-trivial factor of N that can be computed using $O(L^3)$ operations.*

Note that if $x \in [1, N]$, then we have:

$$x \neq 1 \pmod{N} \Rightarrow x \neq 1, \quad \text{and} \quad x \neq -1 \pmod{N} \Rightarrow x \neq N - 1.$$

The problem is thus reduced to find a non-trivial solution x to $x^2 = 1 \pmod{N}$. This second theorem can help us.

Theorem 5.2 *Suppose $N = p_1^{\alpha_1} \ldots p_m^{\alpha_m}$ is the prime factorization of an odd composite positive integer. Let y be an integer chosen uniformly at random, subject to the requirements that $1 \leq y \leq N-1$ and y is co-prime to N, namely $\gcd(y, N) = 1$. Let r be the order of y modulo N, that is the least positive integer such that $y^r \pmod{N} = 1$. Then the probability that r is even and $y^{r/2} \neq -1 \pmod{N}$ satisfies:*

$$p\left(r \text{ even and } y^{r/2} \neq -1 \pmod{N}\right) \geq 1 - \frac{1}{2^m}. \tag{5.49}$$

---

[1] For the proof, see, for instance, G. H. Hardy and E. M. Wright, *An Introduction to the Theory of Numbers* – 6th edition (Oxford University Press, 2008) – Chapter 6.

5.3 The Factoring Algorithm (Shor's Algorithm)

Therefore, the factorizing problem is equivalent to find the order r of random number y modulo N [note that if $y = 1$, its order is $r = 1$, being $1^r (\text{mod } N) = 1$, $\forall r > 0$]: if r is even and $x = y^{r/2}$ is not a trivial solution of $x^2 = 1 (\text{mod } N)$, and this is quite likely according to Theorem 5.2, then we can apply Theorem 5.1, that is, one of $\gcd(x - 1, N)$ and $\gcd(x + 1, N)$ is a non-trivial factor of N.

5.3.1 Order-Finding Protocol

To find the order of $x (\text{mod } N)$ is a *hard problem* on a classical computer, since there is not an algorithm to solve this problem using resources polynomial in $O(L)$, where $L = \lceil \log_2 N \rceil$ is the number of bits needed to specify N. In the following we investigate the performance of a quantum algorithm.

We start from a unitary operator $\hat{U}_x$ such that:

$$\hat{U}_x |y\rangle_L = |xy (\text{mod } N)\rangle_L, \tag{5.50}$$

where $0 \leq y < 2^L$. In Fig. 5.7 we report the quantum circuits representing the action of $\hat{U}_x$. Let us now consider the state:

$$|u_s(x, r)\rangle_L = \frac{1}{\sqrt{r}} \sum_{k=0}^{r-1} \exp\left(-2\pi i \frac{ks}{r}\right) \left|x^k (\text{mod } N)\right\rangle_L \tag{5.51}$$

with $0 < s < r$ integer and r is the (unknown!) order of x modulo N, namely, $x^r (\text{mod } N) = 1$. Note that:

$$_L\langle u_t(x, r)|u_s(x, r)\rangle_L = \delta_{t,s}. \tag{5.52}$$

We have:

$$\hat{U}_x |u_s(x, r)\rangle_L = \frac{1}{\sqrt{r}} \sum_{k=0}^{r-1} \exp\left(-2\pi i \frac{ks}{r}\right) \left|x^{k+1} (\text{mod } N)\right\rangle_L \tag{5.53}$$

$$= \frac{1}{\sqrt{r}} \sum_{k=1}^{r} \exp\left[-2\pi i \frac{(k-1)s}{r}\right] \left|x^k (\text{mod } N)\right\rangle_L \tag{5.54}$$

(a) $|y\rangle_L \dashv\boxed{\hat{U}_x}\vdash |xy(\text{mod } N)\rangle_L$

(b) $|y\rangle_L \dashv\boxed{x(\text{mod } N)}\vdash |xy(\text{mod } N)\rangle_L$

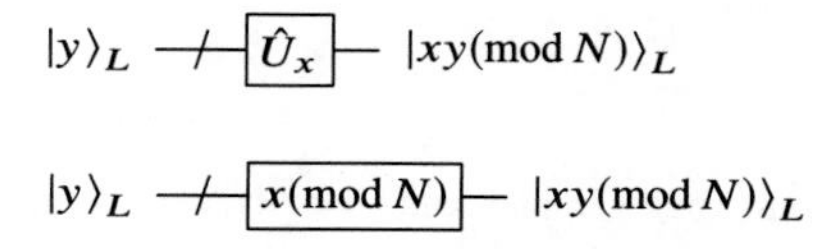

Fig. 5.7 (a) Quantum circuit representing the action of the $\hat{U}_x$ gate acting on the input state $|y\rangle_L$ of L qubits. (b) For the sake of simplicity we can substitute to the symbol $\hat{U}_x$ the expression $x(\text{mod } N)$

$$= \exp\left(2\pi i \frac{s}{r}\right) \frac{1}{\sqrt{r}} \sum_{k=1}^{r} \exp\left(-2\pi i \frac{ks}{r}\right) \left|x^k (\bmod N)\right\rangle_L. \quad (5.55)$$

Since $|x^r (\bmod N)\rangle_L = |x^0 (\bmod N)\rangle_L = |1\rangle$ we can write the last equation as:

$$\hat{U}_x |u_s(x,r)\rangle_L = \exp\left(2\pi i \frac{s}{r}\right) \underbrace{\frac{1}{\sqrt{r}} \sum_{k=0}^{r-1} \exp\left(-2\pi i \frac{ks}{r}\right) \left|x^k (\bmod N)\right\rangle_L}_{|u_s(x,r)\rangle_L} \quad (5.56a)$$

$$= \exp\left(2\pi i \frac{s}{r}\right) |u_s(x,r)\rangle_L \quad (5.56b)$$

$$\equiv \exp\left[2\pi i \, \phi_s(r)\right] |u_s(x,r)\rangle_L \quad (5.56c)$$

where we introduced:

$$\phi_s(r) = \frac{s}{r}. \quad (5.57)$$

It follows that $|u_s(x,r)\rangle_L$ is an eigenstate of $\hat{U}_x$ with eigenvalue $\exp\left(2\pi i \frac{s}{r}\right)$. Therefore, we can estimate $\phi_s(r)$ applying the phase-estimation procedure described in Sect. 5.2. The quantum circuit implementing the order-finding procedure is sketched in Fig. 5.8.

Indeed, we should be able to implement the controlled-$\hat{U}^{2^k}$ gates, and this is fine. The issue could be the preparation of the eigenstate $|u_s(x,r)\rangle_L$. However we note that:

$$\frac{1}{\sqrt{r}} \sum_{s=0}^{r-1} |u_s(x,r)\rangle_L = \frac{1}{r} \sum_{k=0}^{r-1} \sum_{s=0}^{r-1} \underbrace{\exp\left(-2\pi i \frac{sk}{r}\right)}_{r\,\delta_{k,0}} \left|x^k (\bmod N)\right\rangle_L = |1\rangle_L.$$

$$(5.58)$$

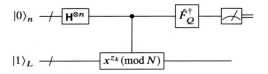

Fig. 5.8 Quantum circuit implementing the order-finding procedure. After the Hadamard transformations the first register is $2^{-n/2} \sum_{z=0}^{2^n-1} |z\rangle_n$, with $|z\rangle_n = |z_n\rangle|z_{n-1}\rangle \ldots |z_0\rangle$

5.3 The Factoring Algorithm (Shor's Algorithm)

Therefore, if we prepare the state $|1\rangle_L \equiv |1(\bmod N)\rangle_L$, we are also preparing a balanced superposition of all the r states $|u_s(x,r)\rangle_L$, $0 \leq s < r$, each with probability $1/r$. Let $1 - \varepsilon$ be the success probability for the estimation of s/r for a given $|u_s(x,r)\rangle$, then the *overall* success probability (we do not know the actual value of s since we have a superposition) is $(1 - \varepsilon)/r$.

Now we investigate how to implement a quantum circuit for the order-finding procedure. As for the usual phase-estimation protocol, we start from the input state $|0\rangle_n|1\rangle_L$ and apply $\mathbf{H}^{\otimes n}$ to the first register, that is to $|0\rangle_n$, obtaining the balanced superposition of all the integers from 0 to $2^n - 1$:

$$\frac{1}{2^{n/2}} \sum_{z=0}^{2^n-1} |z\rangle_n |1\rangle_L. \tag{5.59}$$

We can calculate the action of the controlled $\hat{U}_x^{2^k}$, $k = 0, \ldots, n-1$, on $|1\rangle_L$, where, for a given $|z\rangle_n = |z_{n-1}\rangle \ldots |z_0\rangle$, $z = \sum_{h=0}^{n-1} z_h 2^h$, the control qubit is $|z_k\rangle$. In general we can write:

$$|z\rangle_n |1\rangle_L \longrightarrow |z\rangle_n \frac{1}{\sqrt{r}} \sum_{s=0}^{r-1} \hat{U}_x^{z_{n-1}2^{n-1}} \ldots \hat{U}_x^{z_0 2^0} |u_s(x,r)\rangle_L \tag{5.60a}$$

$$|z\rangle_n \frac{1}{\sqrt{r}} \sum_{s=0}^{r-1} \exp\left[2\pi i \left(z_{n-1} 2^{n-1} + \ldots + z_0 2^0\right) \phi_s(r)\right] |u_s(x,r)\rangle_L \tag{5.60b}$$

$$|z\rangle_n \frac{1}{\sqrt{r}} \sum_{s=0}^{r-1} \exp\left[2\pi i \, z \, \phi_s(r)\right] |u_s(x,r)\rangle_L. \tag{5.60c}$$

Therefore, after the controlled-$\hat{U}_x^{2^k}$ we have the final state (before the inverse of QFT):

$$\frac{1}{\sqrt{r}} \sum_{s=0}^{r-1} \left\{ \frac{1}{2^{n/2}} \sum_{z=0}^{2^n-1} \exp\left[2\pi i \, z \, \phi_s(r)\right] |z\rangle_n \right\} |u_s(x,r)\rangle_L. \tag{5.61}$$

Finally, we can rewrite the state (5.61) as follows:

$$\frac{1}{\sqrt{r}} \sum_{s=0}^{r-1} |\Psi[\phi_s(r)]\rangle_n |u_s(x,r)\rangle_L, \tag{5.62}$$

where:

$$|\Psi[\phi_s(r)]\rangle_n = \frac{1}{2^{n/2}} \sum_{z=0}^{2^n-1} \exp[2\pi i z \phi_s(r)] |z\rangle_n, \quad (5.63)$$

that has the same form as in Eq. (5.29). If we now suppose to measure (implicit measurement) the output register and to find as outcome the state $|u_s(x,r)\rangle_L$ (with probability $1/r$), then the input register is left in the state $|\Psi[\phi_s(r)]\rangle_n$. It is now clear that $\hat{F}_Q^\dagger |\Psi[\phi_s(r)]\rangle_n$ leads to an estimation of $\phi_s(r)$ as shown in the next section.

We have seen how the order-finding problem is reduced to a phase estimation process, where the unknown phase to be estimated is $\phi_s(r) = s/r$. Of course, at the end of the protocol we obtain an estimated value ϕ of $\phi_s(r)$, where both s and r are unknown, thus we should find a way to retrieve this information starting from ϕ. This will be shown in Sect. 5.3.2.

5.3.2 Continued-Fraction Algorithm

First of all we recall that the continued-fraction algorithm describes a positive real number z in terms of positive integers $[a_0, a_1, \ldots, a_M]$, where $a_0 \geq 0$ and $a_k > 0$, $k > 0$, namely:

$$z \to [a_0, a_1, \ldots, a_M] = a_0 + \cfrac{1}{a_1 + \cfrac{1}{\ldots + \cfrac{1}{a_M}}}. \quad (5.64)$$

The m-th convergent to the continued fraction $[a_0, a_1, \ldots, a_M]$ is $[a_0, \ldots, a_m]$, with $0 \leq m \leq M$.

Furthermore, if $z = S/R$, where S and R are $L-bit$ integers, then the algorithm requires $O(L^3)$ operations. For instance, $z = 2.93 \to [2, 1, 13, 3, 2]$. It is also possible decomposing a fraction as a continued fraction, namely

$$z = \frac{31}{13} = 2.\overline{384615} \to [2, 2, 1, 1, 2].$$

In order to find the fraction s/r corresponding to the estimated phase ϕ of $\phi_s(r)$, we can use the following theorem:

Theorem 5.3 *If*

$$\left| \frac{s}{r} - \phi \right| \leq \frac{1}{2r^2} \quad (5.65)$$

then s/r is a convergent of the continued fraction for ϕ and can be computed with $O(L^3)$ operations using the continued-fraction algorithm.

In order to apply the Theorem 5.3 we should satisfy the condition in Eq. (5.65). In our case N is an L-bit integer, and, thanks to the inequality (5.47), the order r is such that $r \leq N \leq 2^L$; we have:

$$\frac{1}{2r^2} \geq \frac{1}{2^{2L+1}}. \tag{5.66}$$

Therefore, if we use $n = 2L + 1$ bits for the register involved in the estimation of $\phi_s(r)$, on the one hand the accuracy in the estimation of the best $\phi^{(b)}$ is $2^{-(2L+1)}$, that is:

$$\left|\phi^{(b)} - \phi\right| \leq \frac{1}{2^{2L+1}}, \tag{5.67}$$

and, on the other hand, Ineqs. (5.66) allow us to write:

$$\left|\phi^{(b)} - \phi\right| \leq \frac{1}{2r^2}, \tag{5.68}$$

and, thus, we can apply the Theorem 5.3 finding the two integers s and r such that:

$$\phi^{(b)} = \frac{s}{r}. \tag{5.69}$$

In particular we obtained the order r and we can check whether $x^r \pmod{N} = 1$.

5.3.3 The Factoring Algorithm

We can now summarize the procedure to factor an integer N:

1. If N is even, return the factor 2.
2. Determine whether $N = a^b$ for integers $a \geq 1$ and $b \geq 2$, and if so return the factor a (this can be done with a classical algorithm).
3. Randomly choose an integer $y \in [1, N-1]$. If $\gcd(y, N) > 1$ then return the factor $\gcd(y, N)$.
4. If $\gcd(y, N) = 1$, use the order-finding subroutine to find the order r of y modulo N (here quantum mechanics help us).
5. If r is even and $x = y^{r/2} \neq -1 \pmod{N}$, then compute $\gcd(x - 1, N)$ and $\gcd(x + 1, N)$, and test to see if one of these is a non-trivial factor N, returning that factor if so (see Theorem 5.1). Otherwise, the algorithm fails.

5.3.4 Example: Factorization of the Number 15

The smallest integer number which is not even or a power of some smaller integer is the number $N = 15$, thus we can apply the order-finding protocol in order to factorize it.

Since $N = 15$, we have $L = \lceil \log_2 15 \rceil = 4$. Therefore, if we require a success probability of at least $1 - \varepsilon = 3/4$, corresponding to an error probability of at most $\varepsilon = 1/4$, the number of qubits needed for the first register is:

$$n = 2L + 1 + \left\lceil \log_2 \left(2 + \frac{1}{2\varepsilon}\right) \right\rceil = 11, \tag{5.70}$$

where the term $2L + 1$ is needed to apply the continued-fraction algorithm (see Sect. 5.3.2).

We proceeds as follows.

1. We generate the random number $y \in [1, N - 1] \equiv [1, 14]$, for instance, we get $y = 7$.
2. We use the order-finding protocol to find the order r of $y \pmod N$. The initial state is $|0\rangle_{11}|1\rangle_4$ and after the application of the Hadamard transformations and the controlled-$\hat{U}^{2^h}$ gates (but before the inverse of the QFT, see Fig. 5.8), we obtain the state:

$$\frac{1}{\sqrt{2048}} \sum_{z=0}^{2047} |z\rangle_{11} |y^z \pmod N\rangle_4, \tag{5.71}$$

which explicitly writes:

$$\frac{1}{\sqrt{2048}} \Big(|0\rangle_{11}|1\rangle_4 + |1\rangle_{11}|7\rangle_4 + |2\rangle_{11}|4\rangle_4 + |3\rangle_{11}|13\rangle_4$$
$$+ |4\rangle_{11}|1\rangle_4 + |5\rangle_{11}|7\rangle_4 + |6\rangle_{11}|4\rangle_4 + |7\rangle_{11}|13\rangle_4 + \ldots \Big), \tag{5.72}$$

or, in a more compact form:

$$\frac{1}{\sqrt{512}} \sum_{k=0}^{511} \frac{1}{2} \big(|4k\rangle_{11}|1\rangle_4 + |1 + 4k\rangle_{11}|7\rangle_4$$
$$+ |2 + 4k\rangle_{11}|4\rangle_4 + |3 + 4k\rangle_{11}|13\rangle_4 \big), \tag{5.73}$$

where we put in evidence four contributions. Now we should apply $\hat{F}_Q^\dagger$ to the first register. However, since the second register does not undergo further transformations, we can assume that it is measured before the application of the

5.3 The Factoring Algorithm (Shor's Algorithm)

inverse of the QFT: this does not affect the success of the protocol but simplifies the theoretical calculations. The measurement outcome will be one of the four possible states $|1\rangle_4$, $|7\rangle_4$, $|4\rangle_4$ or $|13\rangle_4$ with probability 1/4. Suppose we get $|4\rangle_4$, thus the first register is left into the state (similar results follows from the other outcomes):

$$|\Psi[\phi_s(r)]\rangle_{11} = \frac{1}{\sqrt{512}} \sum_{k=0}^{511} |2+4k\rangle_{11}. \tag{5.74}$$

After the inverse of the QFT the previous state of the first register is transformed into the superposition:

$$\hat{F}_Q^\dagger |\Psi[\phi_s(r)]\rangle_{11} = \frac{1}{\sqrt{512}} \sum_{k=0}^{511} \frac{1}{\sqrt{2048}} \sum_{z=0}^{2047} \exp\left(-2\pi i z \frac{2+4k}{2048}\right) |z\rangle_{11}, \tag{5.75}$$

$$= \sum_{z=0}^{2047} c_z |z\rangle_{11}, \tag{5.76}$$

$$= \frac{|0\rangle_{11} - |512\rangle_{11} + |1024\rangle_{11} - |1536\rangle_{11}}{2}, \tag{5.77}$$

where we introduced:

$$c_z = \frac{1}{1024} \sum_{k=0}^{511} \exp\left(-2\pi i z \frac{2+4k}{2048}\right) \tag{5.78}$$

$$= \frac{e^{i\pi z}}{1024} \cos\left(\frac{\pi z}{512}\right) \frac{\sin(\pi z)}{\sin\left(\frac{\pi z}{512}\right)}, \tag{5.79}$$

which is non null only if z is an integer multiple of 512, namely:

$$z = 0, 512, 1024, 1536.$$

Therefore we have:

$$\hat{F}_Q^\dagger |\Psi[\phi_s(r)]\rangle_{11} = \frac{|0\rangle_{11} - |512\rangle_{11} + |1024\rangle_{11} - |1536\rangle_{11}}{2}. \tag{5.80}$$

The measurement on the first register gives with probability 1/4 one of the four states and let's suppose that we obtain $|1536\rangle_{11}$ (similar results are obtained for $|512\rangle_{11}$). Since $2^{11} = 2048$, our outcome leads to the continued-fraction expansion $1536/2048 = 3/4$ and, therefore, the order of $y = 7$ modulo $N = 15$ is $r = 4$ (the denominator of the fraction), which is even!

3. Since the order r is even and $y^{r/2} = 7^2 = 49 \neq 14 \equiv -1 \pmod{15}$, $x = y^{r/2}$ is a solution of $x^2 = 1 \pmod{N}$ and we can apply the Theorem 5.1 obtaining:

$$\gcd(x - 1, N) = \gcd(48, 15) = 3, \tag{5.81a}$$

$$\gcd(x + 1, N) = \gcd(50, 15) = 5. \tag{5.81b}$$

Finally: $15 = 3 \times 5$.

In the other two cases, namely, $|0\rangle_{11}$ and $|1024\rangle_{11}$, the algorithm fails. In fact, if $|0\rangle_{11}$ it is not possible to retrieve the information about r. In the case of $|1024\rangle_{11}$ we have the continued-fraction expansion $1024/2048 = 1/2$, therefore $r = 2$, that is even, $x = y^{r/2} = 7$ but $7^2 \pmod{15} = 4 \neq 1$ and the algorithm fails.

5.4 The RSA Algorithm

To send messages in a secure fashion, a sender can use a "key" to encrypt it. Of course, the receiver should know the key to successfully decrypt the message and understand it. In a typical public-key cryptosystem, the sender's key is public and distinct from the one used to decryption, that is private or secret. Since we have two different keys, one public and one private, this kind of protocol, or algorithm, is called "asymmetric".

One of the most famous asymmetric cryptography algorithm is the RSA, from the surnames of the inventors Ron Rivest, Adi Shamir and Leonard Adleman. As we will see addressing its simplest version, the security of RSA relies on the difficulty of factoring the product of two large prime numbers. This difficulty is due to the hardness of the factoring problem: the time required by best classical algorithm to solve the problem increases exponentially with the number of bit needed to encode the number to be factorized (see Sect. 4.3). As we have seen in the previous sections, the Shor's algorithm may solve it polynomially, posing a security issue.

In its simplest form, the RSA algorithm can be summarized in the following steps.

1. The receiver chooses two prime numbers, say p and q, and multiplies them obtaining the integer $N = p \times q$. If the message to be sent is associated with the integer number z, then $p, q > z$.
2. Given N, p and q, the receiver evaluates the Euler's totient function (5.46), that now explicitly writes:

$$\varphi(N) = (p - 1)(q - 1). \tag{5.82}$$

3. The receiver chooses a number a, called *public exponent*, such that $\gcd(a, N) = 1$.
4. The receiver should now find two relative integers $x > 0$, called *private exponent*, and y, such that

$$x a + y \varphi(N) = 1, \tag{5.83}$$

that is $x a = 1 \ (\mathrm{mod}\ \varphi(N))$.

5. The couple of numbers $\{N, a\}$ is made public and it is used to encode the message, while the receiver uses $\{N, x\}$ to decrypt it.

In practice, given $\{N, a\}$, the sender encodes the message z as follows:

$$m = z^a \ (\mathrm{mod}\ N), \tag{5.84}$$

and m is sent to the receiver. Note that, after the encoding stage, also the sender is no longer able to retrieve the original message! To decrypt the message, the private exponent x as well as N are needed, in fact, due to Eq. (5.83):

$$m^x \ (\mathrm{mod}\ N) = z. \tag{5.85}$$

To be clearer, let us assume that the message to be sent is $z = 5$ and we, as receiver, choose $p = 7$ and $q = 41$, that is $N = 287$, and the Euler function leads to $\varphi(287) = 240$. The reader can easily verify that the following three numbers fulfils the requirements given above: $a = 77$ (public exponent), $x = 53$ (private exponent) and $y = -17$. The sender encodes the message and send us the number $m = 5^{77} \ (\mathrm{mod}\ 287) = 185$ and, once we receive it, we can use the private exponent to retrieve the original message, namely, $185^{53} \ (\mathrm{mod}\ 287) = 5$.

In conclusion, since quantum computers could efficiently factor large integer numbers through the Shor's algorithm, they undermine the RSA algorithm's security.

Problems

5.1 ♣ Given the $\hat{T}$ gate:

$$\hat{T} \to \begin{pmatrix} 1 & 0 \\ 0 & e^{i\pi/4} \end{pmatrix},$$

evaluate the phase shift applying, step by step, the phase estimation protocol with three qubits in the first register (see also Fig. 5.4).

5.2 Prove that given the integers x, y and N, one has:

$$[x(\text{mod } N)]\,[y(\text{mod } N)] = [xy(\text{mod } N)] \,. \tag{5.86}$$

Further Reading

M.A. Nielsen, I.L. Chuang, *Quantum Computation and Quantum Information* (Cambridge University Press, 2010) – Chapter 5

Quantum Search Algorithm

6

Abstract

In this chapter we address the quantum solution to the search problem and, in particular, we discuss the basic aspects of the Grover's algorithm. The geometric interpretation of the algorithm is also given. We illustrate the quantum search on complete graphs through continuous-time quantum walks: in this case the information is encoded onto the vertices of complete graphs and the query is encoded into the Hamiltonian of the system.

6.1 Quantum Search as Standard Computational Process

We focus on the search through a search space of $N = 2^n$ elements, where each element is identified by an integer index $x \in \Omega = \{0, 1, \ldots, N-1\}$ and, thus, by the state $|x\rangle_n$, and we assume that the search has M solutions. We can represent the instance of the search problem by means of a function:

$$f : \{0, 1, \ldots, N-1\} \to \{0, 1\}, \qquad (6.1)$$

such that:

$$f(x) = 0 \Rightarrow x \text{ is not a solution}, \qquad (6.2a)$$

$$f(x) = 1 \Rightarrow x \text{ is a solution}. \qquad (6.2b)$$

Indeed, we also need an oracle able to recognize the solutions to the search problem. As usual, we assume that the oracle acts as follows:

$$|x\rangle_n |q\rangle \xrightarrow{\hat{O}} |x\rangle |q \oplus f(x)\rangle, \qquad (6.3)$$

© The Author(s), under exclusive license to Springer Nature Switzerland AG 2025
S. Olivares, *A Student's Guide to Quantum Computing*, Lecture Notes in Physics 1038, https://doi.org/10.1007/978-3-031-83361-8_6

where $\hat{O}$ is the quantum operator associated with the oracle and $|q\rangle$ is the oracle qubit, $q \in \{0, 1\}$. Note that $|q\rangle \to |\overline{q}\rangle$ only if $f(x) = 1$, namely, only if x is a solution. Due to the linearity, we also have:

$$|x\rangle_n \frac{|0\rangle - |1\rangle}{\sqrt{2}} \xrightarrow{\hat{O}} |x\rangle_n \frac{|0 \oplus f(x)\rangle - |1 \oplus f(x)\rangle}{\sqrt{2}} \equiv (-1)^{f(x)} |x\rangle_n \frac{|0\rangle - |1\rangle}{\sqrt{2}}. \tag{6.4}$$

Since the state of the oracle qubit is left unchanged, we can focus only on the $|x\rangle$. We have:

$$|x\rangle_n \xrightarrow{\hat{O}} |x\rangle_n \text{ if } x \text{ is not a solution,} \tag{6.5a}$$

$$|x\rangle_n \xrightarrow{\hat{O}} -|x\rangle_n \text{ if } x \text{ is a solution,} \tag{6.5b}$$

that is, the oracle marks a solution x to the problem by shifting the phase of the corresponding qubit state $|x\rangle$. It is worth noting that the oracle does not know the solution: it is just able to recognize a solution.

6.2 Quantum Search: The Grover Operator

We start our search procedure with the n qubits prepared in the state $|0\rangle_n$ and, then, we apply n Hadamard transformations in order to generate a superposition of all the possible states:

$$\mathbf{H}^{\otimes n} |0\rangle_n = \frac{1}{2^{n/2}} \sum_{x=0}^{2^n-1} |x\rangle_n \equiv |\psi\rangle_n. \tag{6.6}$$

Now we apply the so-called *Grover iteration* or *Grover operator* $\hat{G}$ which consists in the following steps:

- apply the oracle (this needs also the additional oracle qubit that we do not consider explicitly): $|x\rangle_n \xrightarrow{\hat{O}} (-1)^{f(x)} |x\rangle_n$;
- apply $\mathbf{H}^{\otimes n}$;
- apply the conditional shift $|x\rangle_n \to (-1)^{1+\delta_{x,0}} |x\rangle_n$, i.e., all the states but $|0\rangle_n$, which is left unchanged, undergo a phase shift;
- apply $\mathbf{H}^{\otimes n}$.

Note that the conditional phase shift can be described by the unitary operator $2|0\rangle_n \langle 0| - \hat{\mathbb{I}}$. Furthermore, we have:

$$\mathbf{H}^{\otimes n}(2|0\rangle_n \langle 0| - \hat{\mathbb{I}}) \mathbf{H}^{\otimes n} = 2|\psi\rangle_n \langle \psi| - \hat{\mathbb{I}}, \tag{6.7}$$

therefore, the Grover operator can be written as:

$$\hat{G} = \left[\left(2|\psi\rangle_n\langle\psi| - \hat{\mathbb{I}}\right) \otimes \hat{\mathbb{I}}\right]\hat{O}. \qquad (6.8)$$

The action of the operator $2|\psi\rangle_n\langle\psi| - \hat{\mathbb{I}}$ is also referred to as "inversion by the mean". In fact, given the state:

$$|\phi\rangle_n = \sum_{y=0}^{2^n-1} c_y |y\rangle_n, \qquad (6.9)$$

with $\sum_{y=0}^{2^n-1} |c_y|^2 = 1$, we have:

$$\left(2|\psi\rangle_n\langle\psi| - \hat{\mathbb{I}}\right)|\phi\rangle_n = 2\frac{1}{2^{n/2}}\sum_{y=0}^{2^n-1}\left(\frac{1}{2^{n/2}}\sum_{x=0}^{2^n-1} c_x\right)|y\rangle_n - |\phi\rangle_n$$

$$= \sum_{x=0}^{2^n-1}\left(2\langle c\rangle - c_n\right)|y\rangle_n, \qquad (6.10)$$

where we defined the mean:

$$\langle c\rangle = 2^{-n}\sum_{y=0}^{2^n-1} c_n. \qquad (6.11)$$

In the following we see that by applying $\hat{G}$ a certain number of times, one obtains a solution to the search problem with high probability.

6.2.1 Geometric Interpretation of the Grover Operator

By definition, the state $|\psi\rangle_n$ is a superposition of *all* the possible states $|x\rangle_n$, $x \in \Omega$. However, we can introduce the two sets A and B, $A \cup B = \Omega$ and $A \cap B = \emptyset$, such that:

if $x \in A$ then $f(x) = 0 \Rightarrow x$ is not a solution,

if $x \in B$ then $f(x) = 1 \Rightarrow x$ is a solution.

Therefore we can define the two orthogonal sates:

$$|\alpha\rangle_n = \frac{1}{\sqrt{N-M}} \sum_{x \in A} |x\rangle_n, \quad \text{and} \quad |\beta\rangle_n = \frac{1}{\sqrt{M}} \sum_{w \in B} |w\rangle_n, \qquad (6.12)$$

where $|\alpha\rangle_n$ represents the superposition of all the states $|x\rangle_n$ which are not solutions, while $|\beta\rangle_n$ is the superposition of all the states $|x\rangle_n$ which are solutions to the search problem. Of course we have:

$$|\psi\rangle_n = \sqrt{\frac{N-M}{N}} |\alpha\rangle_n + \sqrt{\frac{M}{N}} |\beta\rangle_n. \qquad (6.13)$$

Since we reduced our N-dimensional system to a two-dimensional one, we can also introduce the following parameterization:

$$|\psi\rangle_n = \cos\frac{\theta}{2} |\alpha\rangle_n + \sin\frac{\theta}{2} |\beta\rangle_n, \qquad (6.14)$$

with:

$$\cos\frac{\theta}{2} = \sqrt{\frac{N-M}{N}}, \quad \text{and} \quad \sin\frac{\theta}{2} = \sqrt{\frac{M}{N}}. \qquad (6.15)$$

We can represent the states $|\alpha\rangle_n$, $|\beta\rangle_n$ and $|\psi\rangle_n$ in a two-dimensional (real) space, as shown in the left panel of Fig. 6.1. This allows us to obtain a geometrical interpretation of the action of the Grover's algorithm. After the query to the oracle we have $|\beta\rangle_n \to -|\beta\rangle_n$, therefore, the state $|\psi\rangle_n$ is reflected across the direction of the vector associated with $|\alpha\rangle_n$ (Fig. 6.1, center panel). Now we should apply $2|\psi\rangle_n\langle\psi| - \hat{\mathbb{I}}$, which corresponds to a reflection across the direction of the vector

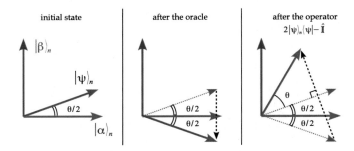

Fig. 6.1 Geometric representation of the action of the Grover operator onto the state $|\psi\rangle_n$ (gray vector): (left) initial state; (center) after the oracle call the initial state is reflected across the direction of the $|\alpha\rangle_n$; (right) after the application of the operator $2|\psi\rangle_n\langle\psi| - \hat{\mathbb{I}}$ the final state is nearer to the vector of solution $|\beta\rangle_n$. The overall effect of a single application of the Grover operator is a counterclockwise rotation of amount θ applied to the initial state $|\psi\rangle_n$

associated with $|\psi\rangle_n$ (right panel of Fig. 6.1). Overall, the action of $\hat{G}$ on $|\psi\rangle_n$ after a single iteration can be summarized as follows (recall that we are not explicitly considering the oracle qubit, which is indeed necessary to apply $\hat{O}$):

$$|\psi\rangle_n = \cos\frac{\theta}{2}|\alpha\rangle_n + \sin\frac{\theta}{2}|\beta\rangle_n \xrightarrow{\hat{G}} \left|\psi^{(1)}\right\rangle_n \quad (6.16)$$

$$\text{with:}\quad \left|\psi^{(1)}\right\rangle_n = \cos\frac{3\theta}{2}|\alpha\rangle_n + \sin\frac{3\theta}{2}|\beta\rangle_n, \quad (6.17)$$

thus, from the geometrical point of view, the action of the Grover operator onto a state is a counterclockwise rotation of an amount θ, described by the matrix:

$$\hat{G} \to \begin{pmatrix} \cos\theta & -\sin\theta \\ \sin\theta & \cos\theta \end{pmatrix}. \quad (6.18)$$

After k iterations we find:

$$|\psi\rangle_n \xrightarrow{\hat{G}^k} \left|\psi^{(k)}\right\rangle_n = \cos\left(\frac{2k+1}{2}\theta\right)|\alpha\rangle_n + \sin\left(\frac{2k+1}{2}\theta\right)|\beta\rangle_n. \quad (6.19)$$

It is worth noting that θ is a function of both N, the total number of states, and of the number of solutions M.

6.2.2 Number of Iterations and Error Probability

As a matter of fact, we have a best number $\mathcal{R}$ of Grover iterations, which bring the initial state $|\psi\rangle_n$ as nearer as possible to the state $|\beta\rangle_n$: further iterations would drive the state away form $|\beta\rangle_n$. Thanks to the geometrical interpretation (see again the left panel of Fig. 6.1) we find that in order to obtain exactly $|\beta\rangle_n$ we should rotate $|\psi\rangle_n$ by an amount $\phi = \arccos\sqrt{M/N}$. Therefore the number of needed iterations is:

$$\mathcal{R} = \text{CI}\left(\frac{\arccos\sqrt{M/N}}{\theta}\right), \quad (6.20)$$

where $\text{CI}(z)$ corresponds to the closest integer to the real number z. After this number of iterations, one measures the final state in the computational basis and obtains a solution to the search problem with a high probability.

In particular, if $M \ll N$, we have that the angular error in the final state will be at most $\theta/2 \approx \sqrt{M/N}$, and the probability of error is thus given by:

$$P_{\text{err}} = \left|\sin\frac{\theta}{2}\right|^2 \approx \frac{M}{N} \ll 1. \quad (6.21)$$

Fig. 6.2 Plot of the r.h.s. of Eq. (6.24)

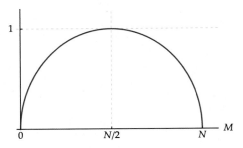

Furthermore, since:

$$\mathcal{R} = \text{CI}\left(\frac{\arccos\sqrt{M/N}}{\theta}\right) \leq \left\lceil \frac{\pi}{2\theta} \right\rceil, \quad (6.22)$$

assuming $M \leq N/2$ we find $\theta/2 \geq \sin(\theta/2) = \sqrt{M/N}$ and we have the following bound on the best number of iterations, i.e.:

$$\mathcal{R} \leq \left\lceil \frac{\pi}{2\theta} \right\rceil \leq \left\lceil \frac{\pi}{4}\sqrt{\frac{N}{M}} \right\rceil, \quad (6.23)$$

that is $\mathcal{R} \sim O(\sqrt{N/M})$, while a classical algorithm would solve the search problem with $O(N)$ steps. It is worth noting that since:

$$\sin\theta = \frac{2\sqrt{M(N-M)}}{N}, \quad (6.24)$$

on the one hand if $M \leq N/2$, then θ grows with the number of solutions M, thus requiring less iterations; on the other hand, if $N/2 < M \leq N$, then θ decreases as M increases, namely, more iterations are required (see Fig. 6.2). This is a silly property of the quantum search algorithm, which can be solved by increasing the total number of states from $N = 2^n$ to $2N = 2^{n+1}$, that is, we just add one qubit.

6.2.3 Quantum Counting

Up to now we addressed the search problem assuming that the number of solutions, and, thus, θ, was known. In general this is not the case. Nevertheless, it is possible to *estimate* both θ and M, and this allows us to find a solution quickly and also to decide whether or not a solution even exists!

In Sect. 6.2.1 we have seen that in the space spanned by $|\alpha\rangle_n$ and $|\beta\rangle_n$, $\hat{G}$ behaves as a rotation described by the 2×2 matrix of Eq. (6.18). It is straightforward to see that $e^{i\theta}$ and $e^{i(2\pi-\theta)}$ are the eigenvalues of $\hat{G}$, therefore we can apply the phase estimation protocol described in Sect. 5.2 in order to estimate θ and M. For ease the analysis, we double N by adding a qubit in order to be assured that the number of solution M is less then the half of the possible states, that is $2N$. Now, we have $\sin^2(\theta/2) = M/(2N)$.

Following Sect. 5.2, if we want an accuracy to m bits, namely, $|\Delta\theta| \leq 2^{-m}$, with success probability $1 - \varepsilon$, we need to use a register with at least a number of qubits given by Eq. (5.43). By using $\sin^2(\theta/2) = M/(2N)$ one can show that:

$$|\Delta M| < \left(2\sqrt{NM} + \frac{N}{2^{m+1}}\right) 2^{-m}. \tag{6.25}$$

6.2.4 Example of Quantum Search

As an example of quantum search we consider a 2-bit search space, that is $N = 2^2$ and we assume to know that there is only one solution to the problem, that is $x_0 \in \{0, 1, 2, 3\}$. From the classical point of view one would need on average 2.25 oracle calls. What is the performance of the quantum algorithm?

We start, as usual, with the superposition:

$$|\psi\rangle_2 = \frac{1}{2} \sum_{x=0}^{3} |x\rangle_2 = \frac{\sqrt{3}}{2} |\alpha\rangle_2 + \frac{1}{2} |\beta\rangle_2, \tag{6.26}$$

where $|\alpha\rangle_2 = 3^{-1/2} \sum_{x \neq x_0} |x\rangle_2$ and $|\beta\rangle_2 = |x_0\rangle_2$. Since $\sin(\theta/2) = 1/2$, we have $\theta = \pi/3$, and, therefore, we need just one iteration of $\hat{G}$ with $\theta = \pi/3$. After the application of the oracle we have:

$$|\psi\rangle_2 \rightarrow \frac{1}{2} \sum_{x \neq x_0} |x\rangle_2 - \frac{1}{2} |x_0\rangle_2 = \sum_{x=0}^{2^n-1} c_x |x\rangle_n \equiv |\phi\rangle_2. \tag{6.27}$$

According to Eq. (6.10), after the "inversion by the mean" we obtain:

$$|\phi\rangle_2 \rightarrow \sum_{x=0}^{3} (2\langle c\rangle - c_x)|x\rangle_2 = |x_0\rangle_2 \tag{6.28}$$

In summary, we have the following overall evolution:

$$|\psi\rangle_2 \xrightarrow{\hat{G}} |x_0\rangle_2. \tag{6.29}$$

We get the right solution with only one oracle call!

6.3 Quantum Search and Unitary Evolution

Suppose that $x_0 \in \{0, 1, \ldots, 2^n - 1\}$ is the label of the only solution. We guess the Hamiltonian which solves the problem of $|\psi\rangle_n$ as initial state and $|x_0\rangle_n$ as solution. Formally, we want a Hamiltonian $\hat{H}$ such that (we use natural units, i.e., $\hbar \to 1$):

$$\exp\left(-i\hat{H}t\right)|\psi\rangle_n = |x_0\rangle_n, \tag{6.30}$$

after a certain time evolution t. As a matter of fact, $\hat{H}$ should depends on both $|\psi\rangle_n$ and $|x_0\rangle_n$. Therefore, the simplest Hamiltonian we can consider is:

$$\hat{H} = |x_0\rangle_n\langle x_0| + |\psi\rangle_n\langle\psi|. \tag{6.31}$$

For the sake of simplicity and to use the qubit formalism, we define the two following orthogonal states:

$$|0\rangle = |x_0\rangle_n, \quad \text{and} \quad |1\rangle = \frac{1}{\sqrt{N-1}} \sum_{x \neq x_0} |x\rangle_n, \tag{6.32}$$

and we write $|\psi\rangle_n = \alpha|0\rangle + \beta|1\rangle$, with $\alpha = \sqrt{(N-1)/N}$ and $\beta = \sqrt{1/N}$. We have:

$$\hat{H} = (\alpha^2 + 1)|0\rangle\langle 0| + \beta^2|1\rangle\langle 1| + \alpha\beta\left(|0\rangle\langle 1| + |1\rangle\langle 0|\right). \tag{6.33}$$

that is:

$$\hat{H} = \hat{\mathbb{I}} + \alpha(\beta\,\hat{\sigma}_x + \alpha\,\hat{\sigma}_z). \tag{6.34}$$

It follows that [see Eq. (2.39)]:

$$\exp\left(-i\hat{H}t\right) = e^{-it}\left[\cos(\alpha t)\,\hat{\mathbb{I}} - i\sin(\alpha t)(\beta\,\hat{\sigma}_x + \alpha\,\hat{\sigma}_z)\right], \tag{6.35}$$

and we find the following evolution (we neglect the overall phase e^{-it}):

$$\exp\left(-i\hat{H}t\right)|\psi\rangle_n = \cos(\alpha t)\,|\psi\rangle_n - i\sin(\alpha t)\,|x_0\rangle_n. \tag{6.36}$$

By choosing $t = \pi/(2\alpha)$ we have, up to an overall phase, $|\psi\rangle_n \to |x_0\rangle_n$.

The Hamiltonian of Eq. (6.34) can be easily simulated using standard methods based on the result known as "Trotter formula":

Theorem Let $\hat{A}$ and $\hat{B}$ be Hermitian operators. Then for any real t we have:

$$\lim_{k \to \infty} \left[\exp\left(i\hat{A}\frac{t}{k}\right) \exp\left(i\hat{B}\frac{t}{k}\right) \right]^k = \exp\left[i\left(\hat{A} + \hat{B}\right)t\right]. \tag{6.37}$$

6.4 Grover's Algorithm and Continuous-Time Quantum Walks

The search problem investigated in the previous sections can be reformulated as a search on a complete graph of N vertices, that is a graph in which each vertex is connected with the other $N - 1$ vertices (see Fig. 6.3). In this case, the vertices are associated with the entries of the search space, namely, $x \to |x\rangle_n$ with $x \in \Omega = \{0, 1, \ldots, N - 1\}$, and the solutions are represented by marked vertices (whose actual positions on the graph are not known). The search is then pursued considering the so-called continuous-time quantum walk in a N-dimensional Hilbert space supported by the vertices of the graph.

In order to describe the dynamics of the quantum walk on the graph G, we should introduce the Laplacian $\boldsymbol{L} = \boldsymbol{A} - \boldsymbol{D}$ of G, where $\boldsymbol{A}$ is the *adjacency matrix* and $\boldsymbol{D}$ is a diagonal matrix such that D_{xx} is the number of edges that are incident to the vertex x, namely, the degree $\deg(x)$ of the vertex x. The adjacency matrix of an undirected graph is defined as

$$A_{x,y} = \begin{cases} 1 & (x, y) \in G, \\ 0 & \text{otherwise}. \end{cases} \tag{6.38}$$

As mentioned above, we associate the state $|x\rangle_n$ with the vertex x, thus the continuous-time quantum walk is defined by introducing the Hamiltonian:

$$\hat{H}_{\text{qw}} = -\gamma \boldsymbol{L}, \tag{6.39}$$

where γ is the jumping rate to an adjacent vertex (for the sake of simplicity we consider $\hbar = 1$). Since here we consider only regular graphs, $\boldsymbol{D}$ is independent of x

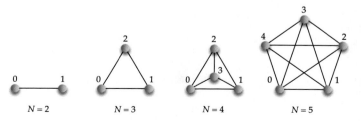

Fig. 6.3 Examples of complete graphs with different values of N. The vertices are represented by the circles while the lines are the edges (the connections between the vertices)

and we can simply assume

$$\hat{H}_{\text{qw}} = -\gamma A. \qquad (6.40)$$

Following the formalism introduced in Sect. 6.2.1, we should introduce the oracle Hamiltonian (the interested reader can find further details in the Further Readings proposed at the end of this chapter)

$$\hat{H}_{\text{sol}} = -\sum_{w \in B} |w\rangle_n \langle w|, \qquad (6.41)$$

$w \in B$ being the solutions, whereas $x \in A$ are the entries which are not solutions, $A \cup B = \Omega$. Note that $\hat{H}_{\text{sol}}$ has eigenvalues equal to zero for all the states but the ground states $|w\rangle_n$, $w \in B$, with eigenvalue -1.

To implement the search on the graph G, we define the Hamiltonian

$$\hat{H} = -\gamma A + \hat{H}_{\text{sol}}, \qquad (6.42)$$

and we consider as initial state $|\psi_0\rangle = |\psi\rangle_n$ given in Eq. (6.6), that is the balanced superposition over the vertices. The evolution of the state at time t is the given by the Schrödinger equation

$$i\frac{\partial}{\partial t}|\psi_t\rangle = \hat{H}|\psi_t\rangle, \qquad (6.43)$$

and the problem is to choose the transition rate γ in such a way that $|\psi_T\rangle$ approaches the superposition of the solution states $|\beta\rangle_n$, introduced in Eq. (6.12), for small a T as possible.

If we consider the Hilbert space spanned by the states $\{|\beta\rangle_n, |\alpha\rangle_n\}$, the Hamiltonian (6.42) can be written in the following matrix form:

$$\hat{H} = -\gamma \begin{pmatrix} M - 1 + \gamma^{-1} & \sqrt{M(N-M)} \\ \sqrt{M(N-M)} & N - M - 1 \end{pmatrix}, \qquad (6.44)$$

where

$$|\beta\rangle_n \to \begin{pmatrix} 1 \\ 0 \end{pmatrix} \quad \text{and} \quad |\alpha\rangle_n \to \begin{pmatrix} 0 \\ 1 \end{pmatrix}. \qquad (6.45)$$

If we now set $\gamma = 1/N$, the eigenstates $\hat{H}|\Psi_\pm\rangle = E_\pm|\Psi_\pm\rangle$ read:

$$|\Psi_\pm\rangle = \frac{|\psi\rangle_n \mp |\beta\rangle_n}{\sqrt{\mathcal{N}_\pm}}, \qquad (6.46)$$

where $|\psi\rangle_n = 2^{-N/2} \sum_{x=0}^{N-1} |x\rangle_n$, with eigenvalues

$$E_\pm = \frac{1-N}{N} \pm \sqrt{\frac{M}{N}}. \tag{6.47}$$

Since from Eq. (6.46) we have

$$|\psi\rangle_n = \frac{\sqrt{N_+}|\Psi_-\rangle + \sqrt{N_-}|\Psi_+\rangle}{\sqrt{2}}, \tag{6.48}$$

by setting $|\psi_0\rangle = |\psi\rangle_n$ we obtain

$$|\psi_t\rangle = e^{-i\hat{H}t}|\psi_0\rangle, \tag{6.49}$$

$$= \frac{\sqrt{N_+}e^{-iE_-t}|\Psi_-\rangle + \sqrt{N_-}e^{-iE_+t}|\Psi_+\rangle}{\sqrt{2}}, \tag{6.50}$$

or, up to a global phase:

$$|\psi_t\rangle = \cos\left(\frac{\Delta E\, t}{2}\right)|\psi\rangle_n - i \sin\left(\frac{\Delta E\, t}{2}\right)|\beta\rangle_n, \tag{6.51}$$

where $\Delta E = E_+ - E_- = 2\sqrt{M/N}$.

It is now clear that if we choose $t = T$ with

$$T = \frac{\pi}{\Delta E} \equiv \frac{\pi}{2}\sqrt{\frac{N}{M}} \tag{6.52}$$

we obtain (up to a global phase) $|\psi_T\rangle = |\beta\rangle_n$. It is worth noting that we obtained the same scaling $\sim O(\sqrt{N/M})$ found in Sect. 6.2.2 in the case of the Grover's algorithm.

Problems

6.1 By using the geometrical representation, prove that $2|\psi\rangle_n\langle\psi| - \hat{\mathbb{I}}$ corresponds to a reflection across the direction of the vector associated with $|\psi\rangle_n$.

6.2 Draw the quantum circuit which implements the quantum search addressed in Sect. 6.2.4.

6.3 ♣ Prove that the Hamiltonian (6.42) can be written in the matrix form (6.44).

Further Readings

A.M. Childs, J. Goldstone, Spatial search by quantum walk. Phys. Rev. A **70**, 022314 (2004)

E. Farhi, S. Gutmann, Analog analogue of a digital quantum computation. Phys. Rev. A **57**, 2403–2406 (1998)

M.A. Nielsen, I.L. Chuang, *Quantum Computation and Quantum Information* (Cambridge University Press, 2010) – Chapter 6

Quantum Operations

7

Abstract

In the real world, a quantum system may be affected by the surrounding environment, and the overall effect is a loss of information and energy leading to decoherence. In this case we should describe the system as an open quantum system and its dynamics is no longer unitary. In this scenario, the quantum operation formalism allows to describe the evolution of the quantum system in a wide variety of circumstances. In the quantum computation context, in particular, it is useful to deal with the (quantum) errors that can occur during a computation. In this chapter we provide some physical and mathematical intuition of the decoherence process and introduce the main noisy maps, that may affect a qubit during the time evolution, namely, the bit flip, the phase flip and the bit-phase flip maps. We also describe other maps that are of general interest: the depolarizing channel, the amplitude damping channel and its generalized version, and the phase damping channel.

7.1 Environment and Quantum Operations

In general, a quantum operation is a map $\mathcal{E}$ that transforms a quantum state described by a density operator $\hat{\varrho}$ into a new density operator $\hat{\varrho}'$, i.e.:

$$\mathcal{E}(\hat{\varrho}) = \hat{\varrho}'. \tag{7.1}$$

A quantum operation captures the dynamic change to a state which occurs as the result of some physical process. The simplest example of quantum operation is the evolution of a quantum state $\hat{\varrho}$ under a unitary operator $\hat{U}$, which can be written as $\mathcal{E}(\hat{\varrho}) \equiv \hat{U} \hat{\varrho} \hat{U}^\dagger$.

According to quantum mechanics, we should also require that the map is convex-linear, namely, given the probabilities $p_k \geq 0$, $\sum_k p_k = 1$, associated with the set of states $\{\hat{\varrho}_k\}$, we have:

$$\mathcal{E}\left(\sum_k p_k \hat{\varrho}_k\right) = \sum_k p_k \mathcal{E}\left(\hat{\varrho}_k\right). \qquad (7.2)$$

Suppose, now, that we have a system S described by $\hat{\varrho}_S$ which interacts with another system E, which we call "environment", described by $\hat{\varrho}_E$. We assume also that the interaction is described by the unitary operator $\hat{U}$. Physically, this corresponds to describe the interaction by means of a Hamiltonian that couples the two systems, leading to their unitary evolution. If S and E are initially uncorrelated, and we are interested just in the evolution of the system, then its evolved state can be represented by the following map:

$$\hat{\varrho}_S \to \mathcal{E}(\hat{\varrho}_S) \equiv \operatorname{Tr}_E\left[\hat{U} \hat{\varrho}_S \otimes \hat{\varrho}_E \hat{U}^\dagger\right]. \qquad (7.3)$$

Without lack of generality we assume that $\hat{\varrho}_E = |e_0\rangle\langle e_0|$, where $\{|e_k\rangle\}$ is an orthonormal basis of the Hilbert space associated with the environment. Now the quantum operation in Eq. (7.3) can be written as:

$$\begin{aligned}
\mathcal{E}(\hat{\varrho}_S) &= \operatorname{Tr}_E\left[\hat{U} \hat{\varrho}_S \otimes |e_0\rangle\langle e_0| \hat{U}^\dagger\right] \\
&= \sum_k \langle e_k| \hat{U} \hat{\varrho}_S \otimes |e_0\rangle\langle e_0| \hat{U}^\dagger |e_k\rangle \\
&= \sum_k \hat{E}_k \hat{\varrho}_S \hat{E}_k^\dagger, \qquad \text{(operator-sum representation)} \qquad (7.4)
\end{aligned}$$

where we introduced $\hat{E}_k = \langle e_k|\hat{U}|e_0\rangle$, called Kraus operators or operation elements, which are linear operators acting on the state space of the system S. Indeed, in order to have a quantum state we should also require that $\forall \hat{\varrho}$, $\operatorname{Tr}_S[\hat{\varrho}] = 1$:

$$\begin{aligned}
1 &= \operatorname{Tr}_S\left[\mathcal{E}(\varrho)\right] \\
&= \operatorname{Tr}_S\left[\sum_k \hat{E}_k \hat{\varrho} \hat{E}_k^\dagger\right] \\
&= \sum_k \operatorname{Tr}_S\left[\hat{E}_k^\dagger \hat{E}_k \hat{\varrho}\right] = \operatorname{Tr}_S\left[\left(\sum_k \hat{E}_k^\dagger \hat{E}_k\right) \hat{\varrho}\right], \qquad (7.5)
\end{aligned}$$

7.1 Environment and Quantum Operations

therefore one should have $\sum_k \hat{E}_k^\dagger \hat{E}_k = \hat{\mathbb{I}}$. More in general one may have $\sum_k \hat{E}_k^\dagger \hat{E}_k \leq \hat{\mathbb{I}}$, and when the inequality is saturated the map is referred to as *trace-preserving*.

Equation (7.3) shows that, a map "summarizes" the effect of a quantum unitary evolution acting not only on the system under consideration. As a matter of fact, there are maps that appears to be "fine" on a single system (they are positive semidefinite, trace preserving and convex-linear), but may be not positive if the addressed system belongs to a larger one. A typical example of this map is the transposition operation described by the map $\mathcal{T}$.

Given the density operator $\hat{\varrho}_A$ of the system A, it is clear that the transposed operator:

$$\mathcal{T}_A(\hat{\varrho}_A) = \hat{\varrho}_A^\mathsf{T} = \hat{\varrho}_A^*, \tag{7.6}$$

is still a density operator describing a physical system (it is self-adjoint, semi-positive definite and with unit trace). However, if:

$$\hat{\varrho}_A = \mathrm{Tr}_B[\hat{\varrho}_{AB}], \tag{7.7}$$

where $\hat{\varrho}_{AB}$ is the state of a larger system, involving the subsystems A and B, the *partially transposed* state:

$$\mathcal{T}_A \otimes \hat{\mathbb{I}}_B(\hat{\varrho}_{AB}), \tag{7.8}$$

could have negative eigenvalues (see, for instance, the problem 7.1).

A map $\mathcal{E}_A$ acting on the density operators $\hat{\varrho}_A \in \mathcal{L}(\mathcal{H}_A)$ of a system A is *positive* if $\mathcal{E}_A(\hat{\varrho}_A) \geq 0$, where $\mathcal{L}(\mathcal{H}_A)$ is the space of the linear operators defined on the Hilbert space $\mathcal{H}_A$. Now, we consider a new arbitrary system B. The map $\mathcal{E}_A$ is *completely positive* (CP) if for any density operator $\hat{\varrho}_{AB} \in \mathcal{L}(\mathcal{H}_A \otimes \mathcal{H}_B)$ one finds:

$$\mathcal{E}_A \otimes \hat{\mathbb{I}}_B(\hat{\varrho}_{AB}) \geq 0. \tag{7.9}$$

Usually, a completely positive and trace preserving map is said to be a CPT map.

The following theorem links CPT map with the operator-sum representation.

Theorem 7.1 *A map* $\mathcal{E}: \mathcal{L}(\mathcal{H}_A) \to \mathcal{L}(\mathcal{H}_B)$ *is trace preserving, convex-linear and completely positive if and only if:*

$$\mathcal{E}(\hat{\varrho}) = \sum_k \hat{E}_k \hat{\varrho} \hat{E}_k^\dagger, \tag{7.10}$$

for some set of operators $\{\hat{E}_k\}$ *mapping the input Hilbert space* $\mathcal{H}_A$ *into the output Hilbert space* $\mathcal{H}_B$ *and such that* $\sum_k \hat{E}_k^\dagger \hat{E}_k = \hat{\mathbb{I}}$.

7.2 Physical Interpretation of Quantum Operations

Suppose we measure the environment in the basis $\{|e_k\rangle\}$. The conditional state $\hat{\varrho}_k$ of the system, corresponding to the outcome k from the measurement, is (we set $\hat{\varrho}_S = \hat{\varrho}$):

$$\hat{\varrho}_k = \frac{1}{p_k} \text{Tr}_E \left[\hat{U}\hat{\varrho} \otimes |e_0\rangle\langle e_0|\hat{U}^\dagger \hat{\mathbb{I}} \otimes \hat{P}_k \right]$$

$$= \frac{1}{p_k} \langle e_k|\hat{U}\hat{\varrho} \otimes |e_0\rangle\langle e_0|\hat{U}^\dagger|e_k\rangle = \frac{1}{p_k} \hat{E}_k\hat{\varrho}\,\hat{E}_k^\dagger, \qquad (7.11)$$

where $\hat{P}_k = |e_k\rangle\langle e_k|$ and:

$$p_k = \text{Tr}_{SE}\left[\hat{U}\hat{\varrho} \otimes |e_0\rangle\langle e_0|\hat{U}^\dagger \hat{\mathbb{I}} \otimes \hat{P}_k \right],$$

$$= \text{Tr}_S \left[\hat{E}_k\hat{\varrho}\,\hat{E}_k^\dagger \right], \qquad (7.12)$$

is the probability of the outcome k. Therefore we have:

$$\mathcal{E}(\hat{\varrho}) = \sum_k \hat{E}_k\hat{\varrho}\,\hat{E}_k^\dagger \equiv \sum_k p_k\hat{\varrho}_k, \qquad (7.13)$$

and the action of $\mathcal{E}$ is to replace $\hat{\varrho}$ with the conditional state $\hat{\varrho}_k$ with probability p_k.

7.3 The Choi–Jamiołkowski Isomorphism

There is a correspondence between quantum channels, characterized by completely positive superoperators, and quantum states, described by density matrices $\hat{\varrho}$, or, more in general, positive operators. This correspondence has been introduced by Man-Duen Choi and Andrzej Jamiołkowski and, sometimes, it is called "channel-state duality", especially in the context of quantum information theory. In reality, we have two different results. On the one hand, there is the Choi isomorphism,[1] on the other hand, we find the Jamiołkowski isomorphism.[2]

Given the CPT map (in reality it is not necessary that the map is also trace preserving):

$$\mathcal{E} : \mathcal{L}(\mathcal{H}_A) \to \mathcal{L}(\mathcal{H}_{A'}), \qquad (7.14)$$

---

[1] M. D. Choi, *Completely positive linear maps on complex matrices*, Linear Algebra Appl. **10**, 285–290 (1975).

[2] A. Jamiołkowski, *Linear transformations which preserve trace and positive semidefiniteness of operators*, Rep. Math. Math. **3**, 275–278 (1972).

7.3 The Choi–Jamiołkowski Isomorphism

we consider an auxiliary system R such that $\dim(\mathcal{H}_A) = \dim(\mathcal{H}_R)$ and we introduce the maximally entangled state (this state can be also not normalized, as the reader will see in the following, but one should check the final normalization of the states):

$$|\Phi_{RA}\rangle = \frac{1}{\sqrt{d}} \sum_{k=0}^{d-1} |k_R\rangle |k_A\rangle. \tag{7.15}$$

Since $\mathcal{E}$ is CPT, the operator:

$$\hat{\eta}_{RA'} = \hat{\mathbb{I}}_R \otimes \mathcal{E}(|\Phi_{RA}\rangle\langle\Phi_{RA}|), \tag{7.16}$$

called "Choi state", is a density operator associated with some state and belonging to $\mathcal{L}(\mathcal{H}_R \otimes \mathcal{H}_{A'})$, and it can be also written as:

$$\hat{\eta}_{RA'} = \sum_{k=0}^{d-1} q_k |\Psi_{RA'}(k)\rangle\langle\Psi_{RA'}(k)|, \tag{7.17}$$

with $q_k \geq 0$ and $\sum_k q_k = 1$.

Due to the linearity of $\mathcal{E}$, given the state:

$$\hat{\varrho}_A = \sum_{j=0}^{d-1} p_j |\varphi_A(j)\rangle\langle\varphi_A(j)|, \tag{7.18}$$

we have:

$$\mathcal{E}(\hat{\varrho}_A) = \sum_{j=0}^{d-1} p_j \mathcal{E}\Big(|\varphi_A(j)\rangle\langle\varphi_A(j)|\Big), \tag{7.19}$$

therefore we can focus only on $\mathcal{E}(|\varphi_A(j)\rangle\langle\varphi_A(j)|)$ and, for the sake of simplicity, we drop the explicit dependence on j.

Since:

$$|\varphi_A\rangle = \sum_{k=0}^{d-1} \varphi_k |k_A\rangle, \tag{7.20}$$

and, from Eq. (7.15):

$$|k_A\rangle = \sqrt{d} \langle k_R|\Phi_{RA}\rangle, \tag{7.21}$$

we can write:

$$|\varphi_A\rangle = \sqrt{d} \sum_{k=0}^{d-1} \langle k_R | \Phi_{RA} \rangle, \tag{7.22}$$

$$= \sqrt{d} \left(\sum_{k=0}^{d-1} \varphi_k \langle k_R| \right) |\Phi_{RA}\rangle, \tag{7.23}$$

$$= \sqrt{d} \, \langle \varphi_R^* | \Phi_{RA} \rangle, \tag{7.24}$$

where:

$$|\varphi_R^*\rangle = \sum_{k=0}^{d-1} \varphi_k^* |k_R\rangle. \tag{7.25}$$

Thereafter, we obtain:

$$\mathcal{E}(|\varphi_A\rangle\langle\varphi_A|) = d\,\mathcal{E}(\langle\varphi_R^*|\Phi_{RA}\rangle\langle\Phi_{RA}|\varphi_R^*\rangle), \tag{7.26}$$

$$= d\,\langle\varphi_R^*|\,\underbrace{\hat{\mathbb{I}}_R \otimes \mathcal{E}(|\Phi_{RA}\rangle\langle\Phi_{RA}|)}_{\hat{\eta}_{RA'}}\,|\varphi_R^*\rangle, \tag{7.27}$$

$$= d\,\mathrm{Tr}_R\big[|\varphi_R^*\rangle\langle\varphi_R^*| \otimes \hat{\mathbb{I}}_{A'}\,\hat{\eta}_{RA'}\big], \tag{7.28}$$

$$= d\,\mathrm{Tr}_R\big[(|\varphi_R\rangle\langle\varphi_R|)^\mathsf{T} \otimes \hat{\mathbb{I}}_{A'}\,\hat{\eta}_{RA'}\big], \tag{7.29}$$

and Eq (7.19) rewrites as:

$$\mathcal{E}(\hat{\varrho}_A) = d\,\mathrm{Tr}_R\big[\hat{\varrho}_R^\mathsf{T} \otimes \hat{\mathbb{I}}_{A'}\,\hat{\eta}_{RA'}\big], \tag{7.30}$$

with:

$$\hat{\varrho}_R = \sum_{j=0}^{d-1} p_j\, |\varphi_R(j)\rangle\langle\varphi_R(j)|. \tag{7.31}$$

Exploiting the expansion (7.17), we eventually find:

$$\mathcal{E}(\hat{\varrho}_A) = d \sum_{k=0}^{d-1} q_k\, \mathrm{Tr}_R\big[\hat{\varrho}_R^\mathsf{T} \otimes \hat{\mathbb{I}}_{A'}\, |\Psi_{RA'}(k)\rangle\langle\Psi_{RA'}(k)|\big], \tag{7.32}$$

$$= d \sum_{k=0}^{d-1} q_k \sum_{j=0}^{d-1} p_j\, \langle\varphi_R^*(j)^*|\Psi_{RA'}(k)\rangle\langle\Psi_{RA'}(k)|\varphi_R^*(j)\rangle, \tag{7.33}$$

7.3 The Choi–Jamiołkowski Isomorphism

$$= \sum_{k=0}^{d-1} \hat{E}_k \hat{\varrho}_A \hat{E}_k^\dagger, \tag{7.34}$$

and we introduced the linear operators $\hat{E}_k : \mathcal{H}_A \to \mathcal{H}_{A'}$ such that:

$$\hat{E}_k |\varphi_A(j)\rangle = \sqrt{dq_k} \, \langle \varphi_R^*(j)^* | \Psi_{RA'}(k) \rangle. \tag{7.35}$$

Note that, if $\mathcal{E}$ is CPT, then $\sum_k \hat{E}_k^\dagger \hat{E}_k = \hat{\mathbb{1}}$.

In summary, starting from the CP of a map $\mathcal{E} : \mathcal{L}(\mathcal{H}_A) \to \mathcal{L}(\mathcal{H}_{A'})$, we mapped a maximally entangled state of the extended Hilbert space $\mathcal{H}_R \otimes \mathcal{H}_A$ to a density operator (the Choi state) on $\mathcal{L}(\mathcal{H}_R) \to \mathcal{L}(\mathcal{H}_{A'})$. The Choi state can be expanded as a convex combination of pure states that can be associated with a Kraus operator of the operator-sum representation of the map $\mathcal{E}$ itself.

Thanks to the isomorphism, one can investigate the properties of a quantum map, or channel, studying the properties of the associated density operator. For instance, a map is completely positive if the corresponding Choi state is positive or a channel is entanglement breaking if the Choi state is separable.

Suppose, now, that we have a unitary operator $\hat{U}$ acting on $\mathcal{H}_A$ and let $\{|k_A\rangle\}$ its basis. Than we can write:

$$\hat{U} = \sum_{j,k} \langle j_A | \hat{U} | k_A \rangle |j_A\rangle\langle k_A|, \tag{7.36}$$

$$= \sum_{j,k} U_{j,k} |j_A\rangle\langle k_A|. \tag{7.37}$$

Since $\hat{U}$ is unitary and, in particular, $\sum_{j,k} |U_{j,k}|^2 = \dim(\mathcal{H}_A)$, it is possible to write the following state (sometimes, it is written as a simple vector, that is without the normalization factor $d^{-1/2}$):

$$|\mathbf{U}\rangle\rangle = \frac{1}{\sqrt{d}} \sum_{j,k} U_{j,k} |j_A\rangle|k_B\rangle, \tag{7.38}$$

where $\mathbf{U}$ is a $d \times d$ matrix, $[\mathbf{U}]_{j,k} = U_{j,k}$, and $|\mathbf{U}\rangle\rangle \in \mathcal{H}_A \otimes \mathcal{H}_B$.

The reader can easily verify that, in the case of the Pauli operators, we have (we drop the subscripts):

$$|\sigma_x\rangle\rangle = \frac{|0\rangle|1\rangle + |1\rangle|0\rangle}{\sqrt{2}}, \tag{7.39}$$

$$|i\sigma_y\rangle\rangle = \frac{|0\rangle|1\rangle - |1\rangle|0\rangle}{\sqrt{2}}, \tag{7.40}$$

$$|\sigma_z\rangle\rangle = \frac{|0\rangle|0\rangle - |1\rangle|1\rangle}{\sqrt{2}}, \tag{7.41}$$

$$|\mathbb{1}\rangle\rangle = \frac{|0\rangle|0\rangle + |1\rangle|1\rangle}{\sqrt{2}}, \tag{7.42}$$

where we also added the identity (note that $i\sigma_y = \sigma_z \sigma_x$).

Before closing this section, we provide a useful application of the isomorphism in the context of bipartite states. Given the generic bipartite state:

$$|\Psi_{AB}\rangle = \sum_{n,m} \psi_{n,m} |n_A\rangle|m_B\rangle, \tag{7.43}$$

we can define the matrix **C** such that $[\mathbf{C}]_{n,m} = \psi_{n,m}$ and, thus, $|\Psi_{AB}\rangle = |\mathbf{C}\rangle\rangle$. Now, we consider two linear operators:

$$\hat{A} = \sum_{k,h} A_{h,k} |k_A\rangle\langle h_A|, \quad \text{and} \quad \hat{B} = \sum_{j,i} B_{j,i} |j_B\rangle\langle i_B|, \tag{7.44}$$

acting on $\mathcal{H}_A$ and $\mathcal{H}_B$, respectively, and the corresponding matrices **A** and **B** defined as usual starting from the matrix elements $A_{h,k} = \langle k_A|\hat{A}|h_A\rangle$ and $B_{j,i} = \langle j_B|\hat{B}|i_B\rangle$. We have:

$$\hat{A} \otimes \hat{B} |\mathbf{C}\rangle\rangle = \sum_{n,m}\sum_{k,h}\sum_{j,i} A_{k,h} B_{j,i} C_{n,m} \langle h_A|n_A\rangle\langle i_B|m_B\rangle |k_A\rangle|j_B\rangle, \tag{7.45}$$

$$= \sum_{k,h}\sum_{j,i} A_{k,h} C_{h,i} B_{j,i} |k_A\rangle|j_B\rangle, \tag{7.46}$$

$$= \sum_{k,j} \left(\sum_{h,i} [\mathbf{A}]_{k,h} [\mathbf{C}]_{h,i} [\mathbf{B}^\mathsf{T}]_{i,j} \right) |k_A\rangle|j_B\rangle, \tag{7.47}$$

$$= \sum_{k,j} [\mathbf{ACB}^\mathsf{T}]_{k,j} |k_A\rangle|j_B\rangle = |\mathbf{ACB}^\mathsf{T}\rangle\rangle. \tag{7.48}$$

7.4 Geometric Picture of Single-Qubit Operations

As we have seen in Chap. 2, we can associate the density operator $\hat{\varrho}$ with a 2×2 density matrix ϱ, which can be written as:

$$\hat{\varrho} \to \varrho = \frac{1}{2}(\mathbb{1} + \mathbf{r} \cdot \boldsymbol{\sigma}) = \frac{1}{2}\begin{pmatrix} 1 + r_z & r_x - ir_y \\ r_x + ir_y & 1 - r_z \end{pmatrix}, \tag{7.49}$$

7.4 Geometric Picture of Single-Qubit Operations

where $\boldsymbol{r} = (r_x, r_y, r_z)$, $\boldsymbol{\sigma} = (\sigma_x, \sigma_y, \sigma_z)$ are the Pauli matrices corresponding to the Pauli operators [see Eqs. (1.27)], and $\boldsymbol{r} \cdot \boldsymbol{\sigma} = r_x \sigma_x + r_y \sigma_y + r_z \sigma_z$. Therefore a trace-preserving quantum operation is equivalent to an affine map of the Bloch sphere into itself and can be written as $\boldsymbol{r} \to \boldsymbol{r}' = \mathbf{M}\boldsymbol{r} + \boldsymbol{v}$, $\mathbf{M}$ being a 3×3 real matrix and $\boldsymbol{v}$ a 3-dimensional real vector.

7.4.1 Bit Flip Operation

If p, with $0 \le p \le 1$, is the probability that a bit flip occurs to a qubit, that is $|0\rangle \to |1\rangle$ and $|1\rangle \to |0\rangle$, the corresponding quantum operation reads:

$$\mathcal{E}_{\text{bf}}(\hat{\varrho}) = (1 - p)\hat{\varrho} + p\hat{\sigma}_x \hat{\varrho} \hat{\sigma}_x, \tag{7.50}$$

and the corresponding elements of the operator-sum representation are:

$$\hat{E}_0 = \sqrt{1 - p}\,\hat{\mathbb{1}}, \quad \text{and} \quad \hat{E}_1 = \sqrt{p}\,\hat{\sigma}_x. \tag{7.51}$$

The transformation of the vector $\boldsymbol{r}$ is (the proof is left to the reader):

$$\begin{cases} r_x \to r_x, \\ r_y \to (1 - 2p)\,r_y, \\ r_z \to (1 - 2p)\,r_z, \end{cases} \tag{7.52}$$

that is we have a contraction of the z–y plane by a factor $1 - 2p$, see Fig. 7.1.

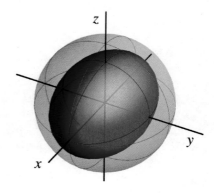

Fig. 7.1 Effect of the bit flip operation on the Bloch sphere: we have a contraction of the z–y plane by a factor $1 - 2p$

Fig. 7.2 Effect of the phase flip operation on the Bloch sphere: we have a contraction of the x–y plane by a factor $1 - 2p$

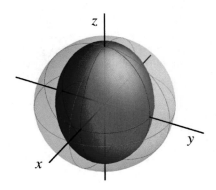

7.4.2 Phase Flip Operation

The quantum operation corresponding to phase flip occurring with probability p is:

$$\mathcal{E}_{\text{pf}}(\hat{\varrho}) = (1 - p)\hat{\varrho} + p\hat{\sigma}_z \hat{\varrho} \hat{\sigma}_z, \tag{7.53}$$

and the corresponding elements of the operator-sum representation are:

$$\hat{E}_0 = \sqrt{1 - p}\,\hat{\mathbb{I}}, \quad \text{and} \quad \hat{E}_1 = \sqrt{p}\,\hat{\sigma}_z. \tag{7.54}$$

The transformation of the vector $\boldsymbol{r}$ is (the proof is left to the reader):

$$\begin{cases} r_x \to (1 - 2p)\, r_x, \\ r_y \to (1 - 2p)\, r_y, \\ r_z \to r_z, \end{cases} \tag{7.55}$$

now we have a contraction of the x–y plane by a factor $1 - 2p$, as shown in Fig. 7.2.

7.4.3 Bit-Phase Flip Operation

When both bit flip and phase flip operations occur with probability p, the process is described by the quantum operation:

$$\mathcal{E}_{\text{bpf}}(\hat{\varrho}) = (1 - p)\hat{\varrho} + p\hat{\sigma}_y \hat{\varrho} \hat{\sigma}_y, \tag{7.56}$$

and the elements of the operator-sum representation are:

$$\hat{E}_0 = \sqrt{1 - p}\,\hat{\mathbb{I}}, \quad \text{and} \quad \hat{E}_1 = \sqrt{p}\,\hat{\sigma}_y. \tag{7.57}$$

7.4 Geometric Picture of Single-Qubit Operations

Fig. 7.3 Effect of the bit-phase flip operation on the Bloch sphere: we have a contraction of the x–z plane by a factor $1 - 2p$

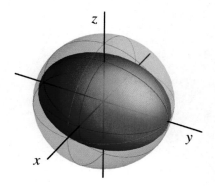

The vector $\boldsymbol{r}$ transforms as follows (the proof is left to the reader):

$$\begin{cases} r_x \to (1-2p)\,r_x, \\ r_y \to r_y, \\ r_z \to (1-2p)\,r_z, \end{cases} \quad (7.58)$$

and, thus, we have a contraction of the x–z plane by a factor $1 - 2p$, see Fig. 7.3.

7.4.4 Depolarizing Channel

The so-called depolarizing channel describes a process in which $\hat{\varrho}$ is replaced by $\hat{\mathbb{I}}/2$, that is the maximally mixed state, with probability p, namely:

$$\mathcal{E}_{\text{dc}}(\hat{\varrho}) = (1-p)\hat{\varrho} + p\frac{\hat{\mathbb{I}}}{2}. \quad (7.59)$$

In order to obtain the operator-sum representation of the depolarizing channel, we use the following identity (the proof is left to the reader):

$$\frac{\hat{\mathbb{I}}}{2} = \frac{1}{4}\left(\hat{\varrho} + \hat{\sigma}_x\,\hat{\varrho}\,\hat{\sigma}_x + \hat{\sigma}_y\,\hat{\varrho}\,\hat{\sigma}_y + \hat{\sigma}_z\,\hat{\varrho}\,\hat{\sigma}_z\right). \quad (7.60)$$

We find:

$$\mathcal{E}_{\text{dc}}(\hat{\varrho}) = \left(1 - \frac{3p}{4}\right)\hat{\varrho} + \frac{p}{4}\sum_{k=x,y,z}\hat{\sigma}_k\,\hat{\varrho}\,\hat{\sigma}_k. \quad (7.61)$$

or:

$$\mathcal{E}_{\text{dc}}(\hat{\varrho}) = (1-q)\,\hat{\varrho} + \frac{q}{3}\sum_{k=x,y,z}\hat{\sigma}_k\,\hat{\varrho}\,\hat{\sigma}_k, \quad (7.62)$$

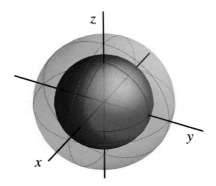

Fig. 7.4 Effect of the depolarizing channel on the Bloch sphere: we have a uniform contraction by a factor $p-1$. The center of the sphere corresponds to the qubit maximally mixed state $\hat{\mathbb{I}}/2$

with $q = 3p/4$, which tells us that the depolarizing channel leaves $\hat{\varrho}$ unchanged with probability $1-q$, while with probability $q/3$ one of the Pauli operators is applied to it. The vector $\boldsymbol{r}$ evolves as follows (the proof is left to the reader):

$$\begin{cases} r_x \to (1-p)\, r_x, \\ r_y \to (1-p)\, r_y, \\ r_z \to (1-p)\, r_z, \end{cases} \quad (7.63)$$

therefore, we have a contraction of the whole sphere by a factor $1-p$. Note that the maximally mixed state, in the Bloch sphere formalism, corresponds to the center of the sphere. Figure 7.4 shows the uniform contraction of the Bloch sphere under the effect of the depolarizing channel.

7.5 Amplitude Damping Channel

Amplitude damping describes the energy dissipation (e.g., an atom which emits a photon, losses during the propagation of light, a system approaching the thermal equilibrium). The map which describes this process is:

$$\mathcal{E}_{\text{ad}}(\hat{\varrho}) = \hat{E}_0 \hat{\varrho} \hat{E}_0^\dagger + \hat{E}_1 \hat{\varrho} \hat{E}_1^\dagger, \quad (7.64)$$

with:

$$\hat{E}_0 = \frac{1}{2}\left[(1+\sqrt{1-\gamma})\,\hat{\mathbb{I}} + (1-\sqrt{1-\gamma})\,\hat{\sigma}_z\right] \to \begin{pmatrix} 1 & 0 \\ 0 & \sqrt{1-\gamma} \end{pmatrix}, \quad (7.65a)$$

$$\hat{E}_1 = \frac{\sqrt{\gamma}}{2}\left(\hat{\sigma}_x + i\hat{\sigma}_y\right) \to \begin{pmatrix} 0 & \sqrt{\gamma} \\ 0 & 0 \end{pmatrix}, \quad (7.65b)$$

$1 \leq \gamma \leq 0$. Note that we can also write $\sqrt{\gamma} = \sin\theta$ and $\sqrt{1-\gamma} = \cos\theta$.

7.5 Amplitude Damping Channel

Since $\hat{E}_0 = |0\rangle\langle 0| + \sqrt{1-\gamma}\,|1\rangle\langle 1|$ and $\hat{E}_1 = \sqrt{\gamma}\,|0\rangle\langle 1|$, it is easy to verify that:

$$\hat{E}_0|0\rangle = |0\rangle, \quad \text{and} \quad \hat{E}_0|1\rangle = \sqrt{1-\gamma}\,|1\rangle, \qquad (7.66)$$

and:

$$\hat{E}_1|0\rangle = 0, \quad \text{and} \quad \hat{E}_1|1\rangle = \sqrt{\gamma}\,|0\rangle, \qquad (7.67)$$

therefore γ can be thought as the probability of loosing a quantum of energy. We have the following effect on the Bloch sphere:

$$\begin{cases} r_x \to \sqrt{1-\gamma}\,r_x, \\ r_y \to \sqrt{1-\gamma}\,r_y, \\ r_z \to \gamma + (1-\gamma)\,r_z. \end{cases} \qquad (7.68)$$

In order to describe the dissipative dynamics affecting a qubit, we make the following substitution:

$$\gamma \to \gamma(t) = 1 - e^{-t/\tau}, \qquad (7.69)$$

where t is a parameter corresponding to the time evolution and τ is a characteristic time of the system (here we assume that $t = 0$ represents the initial time). Inserting $\gamma(t)$ into Eq. (7.64) we obtain a quantum operation describing a dissipative time evolution. In particular, since:

$$\lim_{t \to +\infty} \gamma(t) = 1, \qquad (7.70)$$

as time increases the system evolves toward the state $|0\rangle$ (the north pole of the Bloch sphere), which is the lowest energy level of the qubit: we can now easily understand why the map of Eq. (7.64) represents dissipation ... at least for a quantum system at zero temperature. Figure 7.5 shows the deformation of the Bloch sphere due to the amplitude damping channel (with asymptotic state $\hat{\varrho}_\infty = |0\rangle\langle 0|$).

Fig. 7.5 Effect of the amplitude damping channel on the Bloch sphere with $\hat{\varrho}_\infty = |0\rangle\langle 0|$, that is the north pole of the unit sphere

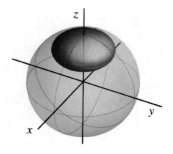

7.6 Generalized Amplitude Damping Channel

In general, quantum systems may have a nonzero temperature T and, in this case, the asymptotic state does not correspond to the lowest energy one. This fact is described by means of a *generalized* amplitude damping channel which involves the two operators $\hat{E}_0$ and $\hat{E}_1$ of Eqs. (7.65) and the following two further operators:

$$\hat{E}_2 = \frac{1}{2}\left[(1+\sqrt{1+\gamma})\hat{\mathbb{I}} - (1-\sqrt{1-\gamma})\hat{\sigma}_z\right] \to \begin{pmatrix} \sqrt{1-\gamma} & 0 \\ 0 & 1 \end{pmatrix}, \quad (7.71a)$$

$$\hat{E}_3 = \frac{\sqrt{\gamma}}{2}(\hat{\sigma}_x - i\hat{\sigma}_y) \to \begin{pmatrix} 0 & 0 \\ \sqrt{\gamma} & 0 \end{pmatrix}, \quad (7.71b)$$

which represent a *phase insensitive* amplification process. In fact, since:

$$\hat{E}_2 = \sqrt{1-\gamma}\,|0\rangle\langle 0| + |1\rangle\langle 1| \quad \text{and} \quad \hat{E}_3 = \sqrt{\gamma}\,|1\rangle\langle 0|, \quad (7.72)$$

it is easy to verify that:

$$\hat{E}_2|0\rangle = \sqrt{1-\gamma}\,|0\rangle, \quad \text{and} \quad \hat{E}_2|1\rangle = |1\rangle, \quad (7.73)$$

and:

$$\hat{E}_3|0\rangle = \sqrt{\gamma}\,|1\rangle, \quad \text{and} \quad \hat{E}_3|1\rangle = 0. \quad (7.74)$$

The whole map reads:

$$\mathcal{E}_{\text{gad}}(\hat{\varrho}) = p\left(\hat{E}_0\hat{\varrho}\hat{E}_0^\dagger + \hat{E}_1\hat{\varrho}\hat{E}_1^\dagger\right) + (1-p)\left(\hat{E}_2\hat{\varrho}\hat{E}_0^\dagger + \hat{E}_3\hat{\varrho}\hat{E}_1^\dagger\right), \quad (7.75)$$

where $0 \le p \le 1$. If we perform the same substitution given in Eq. (7.69), we find that the stationary state for $t \to +\infty$ is:

$$\hat{\varrho}_\infty = \frac{1}{2}\hat{\mathbb{I}} + \frac{2p-1}{2}\hat{\sigma}_z \to \begin{pmatrix} p & 0 \\ 0 & 1-p \end{pmatrix}. \quad (7.76)$$

7.6.1 Approaching the Thermal Equilibrium

When the quantum operation of Eq. (7.75) describes the evolution of a qubit state toward the thermal equilibrium, the probability p is a function of the temperature T. If $\mathcal{E}_x$ is the energy of the state $|x\rangle$, $x = 0, 1$, then one has that the state occupation probability is given by the Boltzmann distribution, namely:

$$p_x(T) = \frac{1}{\mathcal{Z}}\exp\left(-\frac{\mathcal{E}_x}{k_B T}\right), \quad (7.77)$$

where $\mathcal{Z} = p_0(T) + p_1(T)$ is the partition function and k_B is the Boltzmann constant. Therefore the stationary, equilibrium state writes:

$$\hat{\varrho}_\infty(T) \to \begin{pmatrix} p_0(T) & 0 \\ 0 & 1 - p_0(T) \end{pmatrix} \tag{7.78}$$

$$\to \frac{1}{\mathcal{Z}} \begin{pmatrix} \exp[-\mathcal{E}_0/(k_B T)] & 0 \\ 0 & \exp[-\mathcal{E}_1/(k_B T)] \end{pmatrix}, \tag{7.79}$$

which represents the statistical mixture describing a two-level system at thermal equilibrium at temperature T. The purity of the state $\hat{\varrho}_\infty(T)$ is:

$$\mu[\hat{\varrho}_\infty(T)] = 1 - 2 p_0(T) p_1(T). \tag{7.80}$$

7.7 Phase Damping Channel

This kind of channel describes the loss of quantum information without loss of energy. We can derive the quantum operation of this channel addressing a single qubit system subjected to a rotation around the z-axis of the Bloch sphere, namely:

$$\hat{R}_z(\vartheta/2) = \cos\frac{\vartheta}{2}\hat{\mathbb{I}} - i\sin\frac{\vartheta}{2}\hat{\sigma}_z \to \begin{pmatrix} e^{-i\vartheta/2} & 0 \\ 0 & e^{i\vartheta/2} \end{pmatrix}, \tag{7.81}$$

where ϑ is random (this is a random kick). We assume that ϑ is randomly distributed according to a Gaussian distribution with zero mean and variance $2\Delta^2$. We have the following evolution:

$$\hat{\varrho} \to \mathcal{E}_{\text{pdc}}(\hat{\varrho}) = \frac{1}{\sqrt{4\pi\Delta^2}} \int_{-\infty}^{+\infty} d\vartheta \, \exp\left(-\frac{\vartheta^2}{4\Delta^2}\right) \hat{R}_z(\vartheta) \hat{\varrho} \, \hat{R}_z(\vartheta)^\dagger \tag{7.82}$$

$$= \hat{E}_0 \hat{\varrho} \hat{E}_0^\dagger + \hat{E}_0 \hat{\varrho} \hat{E}_0^\dagger, \tag{7.83}$$

with:

$$\hat{E}_0 = \sqrt{\frac{1+\exp(-\Delta^2)}{2}} \hat{\mathbb{I}}, \quad \text{and} \quad \hat{E}_1 = \sqrt{\frac{1-\exp(-\Delta^2)}{2}} \hat{\sigma}_z. \tag{7.84}$$

It is worth noting that the quantum operation of Eq. (7.83) corresponds to the phase flip operation addressed in Sect. 7.4.2 with

$$p = \frac{1+\exp(-\Delta^2)}{2}. \tag{7.85}$$

The effect on the Bloch sphere is analogous to that of the phase flip operation:

$$\begin{cases} r_x \to e^{-\Delta^2} r_x, \\ r_y \to e^{-\Delta^2} r_y, \\ r_z \to r_z. \end{cases} \qquad (7.86)$$

Problems

7.1 ♣ (Partial transposition and Werner states) Given the family of two-qubit states (Werner states):

$$\hat{\varrho}_{AB} = q\,|\Psi_{AB}\rangle\langle\Psi_{AB}| + \frac{1-q}{4}\hat{\mathbb{I}}_{AB}$$

where:

$$|\Psi_{AB}\rangle = \frac{|0_A\rangle|1_B\rangle - |1_A\rangle|0_B\rangle}{\sqrt{2}},$$

and $\hat{\mathbb{I}}_{AB}$ is the identity operator and:

$$q = \frac{3-4p}{3}, \quad p \in [0,1]$$

find the values of p such that the partially transposed state $\mathcal{T}_A \otimes \hat{\mathbb{I}}_B\left(\hat{\varrho}_{AB}\right)$ is not semi-positive definite. Show that if $\mathcal{T}_A \otimes \hat{\mathbb{I}}_B\left(\hat{\varrho}_{AB}\right) < 0$, then $\hat{\varrho}_{AB}$ is entangled.

7.2 Write the amplitude damping map $\mathcal{E}_{\text{ad}}(\hat{\varrho})$ as a function of the Pauli operators.

7.3 Find the evolution of the Bloch vector r under the effect of the generalized amplitude damping channel.

Further Readings

M.A. Nielsen, I.L. Chuang, *Quantum Computation and Quantum Information* (Cambridge University Press, 2010) – Chapter 8.2
J. Preskill, *Lecture Notes for Ph219/CS219: Quantum Information* (2018) – Chapter 3
M.M. Wilde, *Quantum Information Theory* (Cambridge University Press, 2013) – Chapter 4.4

Basics of Quantum Error Correction

8

Abstract

Preserving the information during its processing is of fundamental importance, and this is the main focus of quantum error correction. On the one hand, we should face the problem of errors during the transmission and the storing of quantum information. Here, we show how the three-qubit code (the Shor code) can take care of this issue in the presence of the main errors affecting the qubits (bit flip, phase flip and bit-phase flip). On the other hand, we must avoid the propagation and the accumulation of error due to faulty gates. To this aim, we explain the basic aspects of the fault-tolerant quantum computation and one of its most relevant results: the threshold theorem for quantum computation.

8.1 Quantum Error-Correcting Code and Error Correction Conditions

Given the Hilbert space $\mathcal{H}_{\text{sys}}$ associated with the system we aim at protecting from errors, a quantum error-correcting code is a subspace $\mathcal{C}$ of a larger Hilbert space, $\mathcal{H}_{\text{sys}} \otimes \mathcal{H}_{\text{aux}}$, where $\mathcal{H}_{\text{aux}}$ is an auxiliary Hilbert space. The state of the system we want to protect is then mapped into $\mathcal{C}$ through a suitable unitary operation. Thereafter, the code can be affected by some noise described by a CPT map:

$$\mathcal{E} : \mathcal{L}(\mathcal{C}) \to \mathcal{L}(\mathcal{H}_{\text{sys}} \otimes \mathcal{H}_{\text{aux}}), \tag{8.1}$$

and if $\hat{\varrho} \in \mathcal{L}(\mathcal{C})$ is the density operator associated with the code, the noisy state affected by errors is given by $\mathcal{E}(\hat{\varrho})$.

Now, to correct the errors, first we should diagnose which is the particular error occurred. This can be achieved thanks to a *syndrome measurement*. Once we determined the *error syndrome* (that is the outcome of the syndrome measurement) we can proceed to the *recovery* in order to restore the original state of the code. Usually, the syndrome measurement and the recovery are implemented together and the error correction can be obtained by applying a CPT map $\mathcal{R}$, the error-correction map, such that:

$$(\mathcal{R} \circ \mathcal{E})(\hat{\varrho}) = \hat{\varrho}. \tag{8.2}$$

The syndrome measurement should discriminate among the error syndromes and this is possible if the syndromes map the original code in orthogonal subspaces of $\mathcal{H}_{\text{sys}} \otimes \mathcal{H}_{\text{aux}}$. To this aim, a syndrome measurement operator $\hat{M}_k$, $k = 1, 2, \ldots$, has to be associated with the k-th error syndrome.

A particular quantum error-correcting code protects against a noise $\mathcal{E}$ if some quantum error-correction conditions are satisfied as stated by the following theorem:

Theorem 8.1 (Quantum Error-Correction Conditions) *Let $\mathcal{C}$ be a quantum code and $\hat{P}_\mathcal{C}$ the projector onto $\mathcal{C}$. Given the quantum operation $\mathcal{E}$ with operation elements $\{\hat{E}_k\}$, a necessary and sufficient condition for the existence of an error-correction operation $\mathcal{R}$ correcting $\mathcal{E}$ on $\mathcal{C}$ is that:*

$$\hat{P}_\mathcal{C} \hat{E}_j^\dagger \hat{E}_k \hat{P}_\mathcal{C} = A_{j,k} \hat{P}_\mathcal{C}, \tag{8.3}$$

where $A_{j,k} \in \mathbb{C}$ are the elements of a Hermitian matrix **A**.

In the following we introduce the three-qubit code as an example of the application of quantum error correction.

8.2 The Binary Symmetric Channel

In a classical binary symmetric channel (BSC) the information is encoded into the bits $|0\rangle$ and $|1\rangle$ and we assume that a bit flip error may occur with probability p. The probability of error, that is the probability that $|x\rangle \to |\bar{x}\rangle$, with $x = 0, 1$, is simply given by the bit flip probability, that is:

$$p_{\text{err}}^{(1)} = p, \tag{8.4}$$

where the superscript tell us we are using just one bit to encode the information.

8.2.1 The Three-Bit Code

One of the classical codes used to correct the bit flip error is the three-bit code. Here the information is encoded onto three independent copies of the original bit and the correction strategy is based on the *majority voting*: if, among the received three bits, at least two have the same value x, then we decide that the sent bit value was x.

8.3 Quantum Error Correction: The Three-Qubit Code

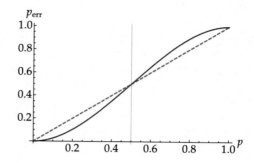

Fig. 8.1 Plot of $p_{\text{err}}^{(1)}$ (dashed, red line) and $p_{\text{err}}^{(3)}$ (solid, blue line) as functions of the bit flip probability p. For values of p less than 0.5 the three-bit code has a better performance with respect to the single bit encoding

Indeed, here we are also assuming that only one bit undergoes bit flip and, thus, we have the following failure probability, which is the probability of having two or more bits flipped:

$$p_{\text{err}}^{(3)} \equiv p_{\geq 2} = p^3 + 3p^2(1-p) = 3p^2 - 2p^3. \tag{8.5}$$

As one can see from Fig. 8.1, we have that $p_{\text{err}}^{(3)} < p_{\text{err}}^{(1)}$ if $p < 1/2$.

8.3 Quantum Error Correction: The Three-Qubit Code

A quantum state cannot be cloned and, since we cannot have three identical copies of an unknown quantum state $|\psi\rangle$ (see Sect. 3.3.1), we cannot directly apply the classical three-bit code to the quantum realm. Furthermore, in contrast to the classical case, we cannot measure the state in order to get information about the error, since the measurement destroys the quantum state... We should find a quantum circuit able to "detect" the eventual error (the bit flip) and to correct it *without destroying the quantum state*. The solution to this problem is given by the three-qubit code, that is the analogous of the classical code.

8.3.1 Correction of Bit Flip Error

As we have seen in Sect. 7.4.1, the evolution of a quantum state $\hat{\varrho} \in \mathcal{L}(\mathcal{H})$ through a bit flip channel can be described by the quantum map:

$$\mathcal{E}(\hat{\varrho}) = (1-p)\,\hat{\varrho} + p\,\hat{\sigma}_x\,\hat{\varrho}\,\hat{\sigma}_x, \tag{8.6}$$

where, now, p is the bit flip probability. In the following we assume that the information is encoded in the qubit state $|\psi\rangle = \alpha|0\rangle + \beta|1\rangle$ and we also have $\hat{\sigma}_x|\psi\rangle = \alpha|1\rangle + \beta|0\rangle$.

Fig. 8.2 This quantum circuit implements the transformation $|\psi\rangle|0\rangle|0\rangle \to \alpha|000\rangle+\beta|111\rangle$, where, in general, $|\psi\rangle = \alpha|0\rangle + \beta|1\rangle$

The basic idea of the three-qubit code is to encode the information onto three qubits as follows:

$$|\psi\rangle \to |\Psi\rangle = \mathbf{C}_{13}\mathbf{C}_{12}|\psi\rangle|0\rangle|0\rangle \qquad (8.7)$$

$$= \alpha|000\rangle + \beta|111\rangle, \qquad (8.8)$$

where, as usual $|xyz\rangle = |x_1\rangle|y_2\rangle|z_3\rangle$, and $\mathbf{C}_{hk}$ is a CNOT operation with the qubit h as control and the qubit $k \neq h$ as target, $h, k = 1, 2, 3$. The reader can verify that this task is obtained by means of the quantum circuit of Fig. 8.2. It is worth noting that $|\Psi\rangle$ is an entangled state. Following Sect. 8.1, the error-correcting code $\mathcal{C} \subset \mathcal{H}^{\otimes 3}$ is the subspace of the three-qubit Hilbert space spanned by the basis vectors $\{|000\rangle, |111\rangle\}$ and the corresponding projector on it is:

$$\hat{P}_\mathcal{C} = |000\rangle\langle 000| + |111\rangle\langle 111|. \qquad (8.9)$$

As in the classical case, we let the bit flip channel affect *independently* each qubit (uncorrelated channels). After the noisy evolution we should implement the error detection, the error syndrome diagnosis and, then, the correction. Here we consider the scenario in which we do not perform the measurement, but we find a way to proceed unitarily. However, it is possible to implement the error detection through a comparative parity-check by adding additional auxiliary qubits.

The overall map $\mathcal{E}^{(3)}$ affecting the code can be written as:

$$\mathcal{E}^{(3)}(\hat{\varrho}) = \left(\mathcal{E} \otimes \mathcal{E} \otimes \mathcal{E}\right)(\hat{\varrho}), \qquad (8.10)$$

$$= (1-p)^3 \hat{\varrho}$$

$$+ (1-p)^2 p \left(\hat{\chi}_1 \hat{\varrho} \hat{\chi}_1^\dagger + \hat{\chi}_2 \hat{\varrho} \hat{\chi}_2^\dagger + \hat{\chi}_3 \hat{\varrho} \hat{\chi}_3^\dagger\right)$$

$$+ (1-p) p^2 \left(\hat{\chi}_1 \hat{\chi}_2 \hat{\varrho} \hat{\chi}_2^\dagger \hat{\chi}_1^\dagger + \hat{\chi}_1 \hat{\chi}_3 \hat{\varrho} \hat{\chi}_3^\dagger \hat{\chi}_1^\dagger + \hat{\chi}_2 \hat{\chi}_3 \hat{\varrho} \hat{\chi}_3^\dagger \hat{\chi}_2^\dagger\right)$$

$$+ p^3 \hat{\chi}_1 \hat{\chi}_2 \hat{\chi}_3 \hat{\varrho} \hat{\chi}_3^\dagger \hat{\chi}_2^\dagger \hat{\chi}_1^\dagger, \qquad (8.11)$$

8.3 Quantum Error Correction: The Three-Qubit Code

where, now, $\hat{\varrho} \in \mathcal{L}(\mathcal{C})$, and we defined:

$$\hat{\chi}_1 = \hat{\sigma}_x \otimes \hat{\mathbb{I}} \otimes \hat{\mathbb{I}}, \quad \hat{\chi}_2 = \hat{\mathbb{I}} \otimes \hat{\sigma}_x \otimes \hat{\mathbb{I}}, \quad \hat{\chi}_3 = \hat{\mathbb{I}} \otimes \hat{\mathbb{I}} \otimes \hat{\sigma}_x. \tag{8.12}$$

In our analysis we still assume that at most one qubit is flipped since, in general, $p \ll 1$, thus we obtain the same failure probability given by Eq. (8.5). In this case, we can consider only the terms proportional to p, and Eq. (8.11) reduces to:

$$\mathcal{E}^{(3)}(\hat{\varrho}) \approx (1 - 3p)\,\hat{\varrho} + p\left(\hat{\chi}_1 \hat{\varrho} \hat{\chi}_1^\dagger + \hat{\chi}_2 \hat{\varrho} \hat{\chi}_2^\dagger + \hat{\chi}_3 \hat{\varrho} \hat{\chi}_3^\dagger\right). \quad (p \ll 1) \tag{8.13}$$

The map is clearly CPT with operation elements:

$$\hat{E}_0 = \sqrt{1 - 3p}\,\hat{\mathbb{I}}, \quad \text{and} \quad \hat{E}_k = \sqrt{p}\,\hat{\chi}_k, \tag{8.14}$$

with $k = 1, 2, 3$. The reader can easily verify that the conditions of Theorem 8.1 are satisfied and the error syndrome measurement operators write:

$$\hat{M}_0 = \hat{P}_\mathcal{C}, \quad \text{and} \quad \hat{M}_k = \hat{\chi}_k\,\hat{P}_\mathcal{C}\,\hat{\chi}_k^\dagger, \quad (k = 1, 2, 3) \tag{8.15}$$

leading to the four possible outcomes (the error syndromes). Otherwise, the error syndromes can be retrieved considering a comparative parity-check measurement between qubits 1–2 and 2–3, described by the operators:

$$\hat{M}_{12}^{(pc)} = \hat{\sigma}_z \otimes \hat{\sigma}_z \otimes \hat{\mathbb{I}}, \quad \text{and} \quad \hat{M}_{23}^{(pc)} = \hat{\mathbb{I}} \otimes \hat{\sigma}_z \otimes \hat{\sigma}_z. \tag{8.16}$$

This measurement leaves the state unchanged but allows to detect the error syndrome. In fact, we have the following possible outcomes depending on the qubit $k = 1, 2, 3$ affected by the error, as summarized in Table 8.1. This kind of scheme requires to add some CNOT gates acting on two ancillary qubits and using the code qubits as control: in this way the parity-check measurement leaves the code unperturbed.

Up to now we have considered the syndrome measurement and the recovery as two different steps of the error correction. Nevertheless, it is possible joining them

Table 8.1 Error correction with comparative parity-check measurement

Flipped qubit	$\hat{M}_{12}^{(pc)}$	$\hat{M}_{23}^{(pc)}$	Recovery operation
No error	$+1$	$+1$	$\hat{\mathbb{I}} \otimes \hat{\mathbb{I}} \otimes \hat{\mathbb{I}}$
Qubit 1	-1	$+1$	$\hat{\sigma}_x \otimes \hat{\mathbb{I}} \otimes \hat{\mathbb{I}}$
Qubit 2	-1	-1	$\hat{\mathbb{I}} \otimes \hat{\sigma}_x \otimes \hat{\mathbb{I}}$
Qubit 3	$+1$	-1	$\hat{\mathbb{I}} \otimes \hat{\mathbb{I}} \otimes \hat{\sigma}_x$

together. When the error occurs on the qubit $k = 1, 2, 3$, the state $|\Psi\rangle$ given in Eq. (8.8) evolves with probability p as:

$$|\Psi\rangle \to |\tilde{\Psi}\rangle = \hat{\chi}_k |\Psi\rangle, \tag{8.17}$$

and it is easy to verify that if $k \neq 1$:

$$(\mathbf{C}_{13}\mathbf{C}_{12})^\dagger \hat{\chi}_k \mathbf{C}_{13}\mathbf{C}_{12} = \hat{\chi}_k, \quad (k \neq 1) \tag{8.18}$$

while for $k = 1$:

$$(\mathbf{C}_{13}\mathbf{C}_{12})^\dagger \hat{\chi}_1 \mathbf{C}_{13}\mathbf{C}_{12} = \hat{\sigma}_x \otimes \hat{\sigma}_x \otimes \hat{\sigma}_x, \tag{8.19}$$

In summary, if the bit flip affects the first qubit ($k = 1$), we have:

$$|\psi\rangle|0\rangle|0\rangle \to (\hat{\sigma}_x|\psi\rangle)|1\rangle|1\rangle, \tag{8.20}$$

and we can correct the error through a Toffoli gate taking the second and the third qubits as controls and the first one as target; in the other cases ($k \neq 1$) we find:

$$|\psi\rangle|0\rangle|0\rangle \to |\psi\rangle|x\rangle|y\rangle, \tag{8.21}$$

that is the first qubit is left in its initial state $|\psi\rangle$ but the final state of the other qubits is flipped if the error changed their state. Nevertheless, in the latter scenario, we note that $x \neq y$ or $x = y = 0$ (if no error occurred), therefore the presence of a final Toffoli gate will not change the state of qubit 1. In Fig. 8.3 we can see the quantum circuit achieving this goal.

In order to better understand how the three-qubit code works in practice, let us assume that after the bit flip channel the state is:

$$|\Psi'\rangle = \hat{\chi}_1 |\Psi\rangle, \tag{8.22}$$

i.e., the first qubit has been flipped. The first CNOT gate, $\mathbf{C}_{13}$, performs the following transformation:

$$|\Psi'\rangle = \alpha|100\rangle + \beta|011\rangle \to \alpha|101\rangle + \beta|011\rangle, \tag{8.23}$$

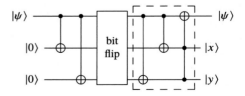

Fig. 8.3 The dashed box encloses the quantum circuit implementing the three-qubit code for quantum error correction against single bit flip operation

8.3 Quantum Error Correction: The Three-Qubit Code

thereafter, we have the second CNOT gate, $\mathbf{C}_{12}$, which leads to:

$$\alpha|101\rangle + \beta|011\rangle \rightarrow \alpha|111\rangle + \beta|011\rangle. \tag{8.24}$$

The last gate is the Toffoli gate, which takes the second and third qubits as control and the first qubit as target, and we get:

$$\alpha|111\rangle + \beta|011\rangle \rightarrow \alpha|011\rangle + \beta|111\rangle \equiv \underbrace{(\alpha|0\rangle + \beta|1\rangle)}_{|\psi\rangle}|11\rangle. \tag{8.25}$$

We conclude that the error has been corrected, the state of the first qubit being $|\psi\rangle$.

As noted above, the code may fail if more than one qubit is flipped. Since the probability that at most one bit is flipped reads:

$$p_{\leq 1} = (1-p)^3 + 3p(1-p)^2 = (1-p)^2(1+2p), \tag{8.26}$$

we have the following probability of error at the output:

$$p_{\text{err, Q}}^{(3)} = 1 - p_{\leq 1} = 3p^2 - 2p^3, \tag{8.27}$$

the same obtained in the classical three-bit code, see Eq. (8.5).

8.3.2 Correction of Phase Flip Error

Phase flip error does not have classical analogue, since the transformation:

$$|1\rangle \rightarrow e^{i\pi}|1\rangle = -|1\rangle, \tag{8.28}$$

does not exist in classical logic. The quantum map describing a channel in which phase flip occurs with probability p reads (see also Sect. 7.4.2):

$$\mathcal{E}(\hat{\varrho}) = (1-p)\hat{\varrho} + p\hat{\sigma}_z \hat{\varrho} \hat{\sigma}_z. \tag{8.29}$$

It is worth noting that since $\hat{\sigma}_z|x\rangle = (-1)^x|x\rangle$, we have:

$$\hat{\sigma}_z|\pm\rangle = |\mp\rangle, \tag{8.30}$$

where:

$$|\pm\rangle = \frac{|0\rangle \pm |1\rangle}{\sqrt{2}}, \tag{8.31}$$

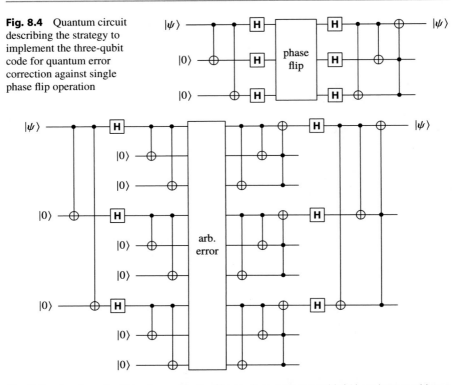

Fig. 8.4 Quantum circuit describing the strategy to implement the three-qubit code for quantum error correction against single phase flip operation

Fig. 8.5 Quantum circuit implementing the Shor code to protect a qubit $|\psi\rangle$ against an arbitrary error

and we conclude that the phase flip channel acts as a bit flip channel on the basis $|\pm\rangle$. Therefore, recalling the action of the Hadamard transformation on the computational basis $|0\rangle$ and $|1\rangle$, it is easy to prove that the quantum circuit represented in Fig. 8.4 corrects a single phase flip error. Actually, the first Hadamard transformations physically change the computational basis in order that the phase flip channel behaves like a bit flip channel; the second Hadamard transformations transform back to the original basis in order to apply the same correction code described in the previous section.

8.3.3 Correction of Any Error: The Shor Code

In a realistic channel both bit and phase flip errors may take place. It is possible to protect the qubit against the effects of an *arbitrary* error by means of the Shor code, which is a combination of the three-qubit bit flip and phase flip error correction codes. In Fig. 8.5 we sketched the quantum circuit implementing the Shor code.

The reader can investigate its action applying the results obtained in the previous sections.

8.4 Fault-Tolerant Quantum Computation

In the previous sections we have described a method to protect stored of transmitted quantum information. However, quantum error-correction is a powerful tool to preserve the information processing during the computation, as well. In the latter case, one remarkable result, known as "quantum fault-tolerance theorem" or, simply, "threshold theorem", states that also in the presence of noisy quantum gates it is possible to achieve arbitrary good quantum computation if the error probability per gate is below a given threshold. Here, the main idea is to prevent the propagation and the accumulation of errors due to the faulty gates.

In general, the noise may affect the preparation and the transmission of the qubits but also the quantum gates and the measurement stage. A possible solution is to replace the *single qubits* with *encoded block of qubits* exploiting a suitable *error-correcting code*, such as the seven-qubit Stean code, belonging to the class of stabilizer codes. In brief, a stabilizer code is based on the stabilizer subgroup of the Pauli group. Such a code exploits the measurement on auxiliary qubits (the stabilizer qubits) to detect errors and, then, to act accordingly on the data qubits. In this scenario, the original gates are replaced with a suitable procedure, the *fault-tolerant procedure*, performing *encoded gates* which act on the encoded state. Error correction is then periodically performed on the encoded state to prevent the accumulation and propagation of the errors. Remarkably, fault-tolerant procedures can be used to implement a universal set of gates, as $\{\mathbf{H}, \text{CNOT}, \hat{S}, \hat{T}\}$.

It is important to design the encoded gates: possible errors should affect only a limited number of qubits in order to make error correction more effective. Moreover, also the error correction should be designed to avoid the introduction of too many errors.

One can show that if p is the probability of failure of an individual component in a quantum circuit implementing a quantum computation, performing any operation in a fault-tolerant fashion allows reducing the error probability from p to cp^2, where c is a constant (usually, $c \approx 10^4$).

The error rate of the computation can be further reduced using concatenated codes, namely, recursively applying encoded circuits. For instance, suppose to use a three-qubit code to encode a single qubit. In the presence of a two-level concatenated code, we have nine qubits (the original qubit is mapped into three qubits and each of those is mapped into other three qubits). In this way the failure probability is reduced from p to cp^2 and, then, to $c(cp^2)^2$, and so on. If we concatenate k times, the failure probability becomes $(cp)^{2^k}/c$.

8.4.1 The Threshold Theorem

We assume to have a problem of size n and the corresponding quantum circuit with a number of gates given by the polynomial $\mathsf{p}(n)$ (think of the factoring algorithm). If we want to simulate fault-tolerantly the circuit with an accuracy ε, each gate should be simulated with accuracy $\varepsilon/\mathsf{p}(n)$. Therefore, the number of concatenations k should satisfy:

$$\frac{(cp)^{2^k}}{c} \leq \frac{\varepsilon}{\mathsf{p}(n)}, \tag{8.32}$$

where p is the failure probability as introduced above. If:

$$p < \frac{1}{c} \equiv p_{\text{th}}, \tag{8.33}$$

we can find k. The threshold condition $p < p_{\text{th}}$, when satisfied, guarantees that the quantum computation can be achieved with arbitrary accuracy.

The following theorem tells us the number of gates needed to simulate a quantum circuit with the desired level of accuracy.

Theorem 8.2 (Threshold Theorem for Quantum Computation) *A quantum circuit containing $\mathsf{p}(n)$ gates may be simulated with probability of error at most ε using:*

$$O\left(\text{poly}\left(\log \mathsf{p}(n)/\varepsilon\right)\mathsf{p}(n)\right) \tag{8.34}$$

gates on hardware whose components fail with probability at most p, provided p is below some constant threshold, $p < p_{\text{th}}$, and given a reasonable assumption about the noise in the underlying hardware.

Here, poly(x) is a polynomial of fixed degree. A rough estimate of threshold probability in the case of the Stean code, where $c \approx 10^4$, leads to $p_{\text{th}} \approx 10^{-4}$, but more accurate and sophisticated calculations yield values between 10^{-5} and 10^{-6}.

Problems

8.1 Prove that:

$$(\mathbf{C}_{13}\mathbf{C}_{12})^\dagger \left(\hat{\mathbb{I}} \otimes \hat{\sigma}_x \otimes \hat{\mathbb{I}}\right) \mathbf{C}_{13}\mathbf{C}_{12} = \hat{\mathbb{I}} \otimes \hat{\sigma}_x \otimes \hat{\mathbb{I}},$$

and that:

$$(\mathbf{C}_{13}\mathbf{C}_{12})^\dagger \left(\hat{\sigma}_x \otimes \hat{\mathbb{I}} \otimes \hat{\mathbb{I}}\right) \mathbf{C}_{13}\mathbf{C}_{12} = \hat{\sigma}_x \otimes \hat{\sigma}_x \otimes \hat{\sigma}_x.$$

8.2 Following, step by step, the action of each gates, verify that the three-qubit circuit depicted in Fig. 8.3 works as follows:

$$\hat{\mathbb{I}} \otimes \hat{\mathbb{I}} \otimes \hat{\mathbb{I}} |\Psi\rangle \to |\psi\rangle|00\rangle,$$

$$\hat{\sigma}_x \otimes \hat{\mathbb{I}} \otimes \hat{\mathbb{I}} |\Psi\rangle \to |\psi\rangle|11\rangle,$$

$$\hat{\mathbb{I}} \otimes \hat{\sigma}_x \otimes \hat{\mathbb{I}} |\Psi\rangle \to |\psi\rangle|10\rangle,$$

$$\hat{\mathbb{I}} \otimes \hat{\mathbb{I}} \otimes \hat{\sigma}_x |\Psi\rangle \to |\psi\rangle|01\rangle.$$

Further Reading

M.A. Nielsen, I.L. Chuang, *Quantum Computation and Quantum Information* (Cambridge University Press, Cambridge, 2010). Chapter 10

Two-Level Systems and Photonic Qubits

Abstract

Any two-level quantum system is associated with a Hilbert space spanned by two orthonormal states and, thus, can be seen as a qubit. In this chapter we will focus on spin–1/2 particles and two-level atoms, which are the simplest example of qubits, and on the degree of freedom of single-photon states. We also explain how it is possible to manipulate spins and atoms in order to implement quantum logic gates.

9.1 Universal Computation with Spins

A typical two-level system is a spin–1/2 particle, usually a nucleus, which can be used as a qubit and manipulated by means of electromagnetic fields.

9.1.1 Interaction Between a Spin–1/2 Particle and a Magnetic Field

The operator associated with the spin magnetic moment of a spin–1/2 particle is given by:

$$\hat{\mu} = -\frac{gq}{2m}\hat{S}, \tag{9.1}$$

where g is the gyromagnetic factor (for an electron $g \approx 2.002$), q and m are the charge and the mass of the particle, respectively, and

$$\hat{S} = \frac{\hbar}{2}\hat{\sigma}, \tag{9.2}$$

where $\hat{\sigma} = (\hat{\sigma}_x, \hat{\sigma}_y, \hat{\sigma}_z)$ is, as usual, the vector of the Pauli operators.

The Hamiltonian describing the interaction between the spin–1/2 particle and the (classical) static magnetic field $\boldsymbol{B} = (B_x, B_y, B_z)$ is:

$$\hat{H}_{\text{int}} = -\hat{\boldsymbol{\mu}} \cdot \boldsymbol{B} = \frac{gq}{2m}\frac{\hbar}{2}\hat{\boldsymbol{\sigma}} \cdot \boldsymbol{B}. \tag{9.3}$$

which can be written as:

$$\hat{H}_{\text{int}} = \frac{\hbar\omega_L}{2}\boldsymbol{n} \cdot \hat{\boldsymbol{\sigma}}, \tag{9.4}$$

where we introduced the Larmor frequency:

$$\omega_L = \frac{gq}{2m}|\boldsymbol{B}|, \tag{9.5}$$

and $\boldsymbol{n} = \boldsymbol{B}/|\boldsymbol{B}|$.

Without lack of generality, we assume $\boldsymbol{B} = (0, 0, B)$, that is we take the magnetic field along the z-direction and, accordingly, $\boldsymbol{n} \cdot \hat{\boldsymbol{\sigma}} = \hat{\sigma}_z$. Given the initial state (as we mentioned, any two-level system can be considered as a qubit, see Sect. 2.2):

$$|\psi_0\rangle = \cos\frac{\theta}{2}|0\rangle + \sin\frac{\theta}{2}|1\rangle, \tag{9.6}$$

with $\hat{\sigma}_z|x\rangle = (-1)^x|x\rangle$, $x = 0, 1$, we have the following time evolution under the effect of $\hat{H}_{\text{int}}$:

$$\begin{aligned}|\psi_t\rangle &= \exp\left(-i\frac{\hat{H}_{\text{int}}}{\hbar}t\right)|\psi_0\rangle, \\ &= \cos\frac{\theta}{2}e^{-i\omega_L t/2}|0\rangle + \sin\frac{\theta}{2}e^{i\omega_L t/2}|1\rangle, \\ &= e^{-i\omega_L t/2}\left(\cos\frac{\theta}{2}|0\rangle + \sin\frac{\theta}{2}e^{i\omega_L t}|1\rangle\right), \end{aligned} \tag{9.7}$$

where, in the last equation, the overall phase $e^{-i\omega_L t/2}$ can be neglected. Following Sect. 2.2.1, the Bloch vector $\boldsymbol{r}_t$ associated with $|\psi_t\rangle$ reads:

$$\boldsymbol{r}_t = \begin{pmatrix} \sin\theta\,\cos(\omega_L t) \\ \sin\theta\,\sin(\omega_L t) \\ \cos\theta \end{pmatrix}, \tag{9.8}$$

that is we have the Larmor precession of the spin around the direction of the magnetic field (here the z-direction), as illustrated in Fig. 9.1.

9.1 Universal Computation with Spins

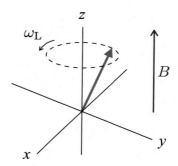

Fig. 9.1 Precession of a spin (red arrow) under the effect of a magnetic field B directed along z-direction. The tip of the arrow representing the spin rotates counterclockwise around the z-direction

More in general, the unitary evolution operator associated with the Hamiltonian (9.4) reads:

$$\exp\left(-i\frac{\hat{H}_{\text{int}}}{\hbar}t\right) = \cos\left(\frac{\omega_L t}{2}\right)\hat{\mathbb{1}} - i\sin\left(\frac{\omega_L t}{2}\right)\boldsymbol{n}\cdot\hat{\boldsymbol{\sigma}}, \quad (9.9)$$

and we can implement single qubit gates by suitably choosing the time t and the amplitude and orientation of the magnetic field $\boldsymbol{B}$.

9.1.2 Spin Qubit and Hadamard Transformation

If we orient the magnetic field along the x-z direction, i.e., $\boldsymbol{B} = B\,\boldsymbol{n}$ with

$$\boldsymbol{n} = \frac{1}{\sqrt{2}}(1, 0, 1), \quad (9.10)$$

and set the evolution time such that $\omega t = \pi$, form Eq. (9.9) we have:

$$\exp\left(-i\frac{\hat{H}_{\text{int}}}{\hbar}t\right) \rightarrow -\frac{i}{\sqrt{2}}(\hat{\sigma}_x + \hat{\sigma}_z) \equiv -i\,\mathbf{H}, \quad (9.11)$$

that is, up to an overall phase factor "$-i$", we have the quantum operator describing the action of the Hadamard transformation introduced in Sect. 1.4.4 [see Eq. (1.31)].

9.1.3 Manipulation of Single Qubit: Nuclear Magnetic Resonance

As a matter of fact, changing the direction of the magnetic field could be not physically possible, since it may require to mechanically move its sources. Fortunately, we can sort to the nuclear magnetic resonance (NMR).

In a typical NMR experiment involving atomic species with nuclear spin of 1/2, the spin interacts with two magnetic fields: the one, $\boldsymbol{B}_0$, is stationary along the z-direction, the other, $\boldsymbol{B}_1(\omega_{RF})$, is a radio-frequency field laying in a plane orthogonal to $\boldsymbol{B}_0$. Without lack of generality we can assume $\boldsymbol{B}_1(\omega_{RF})$ along x-direction with oscillating amplitude $2B_1 \cos(\omega_{RF} t)$ and the Hamiltonian of the system writes:

$$\hat{H}_{NMR} = \frac{\hbar\omega_0}{2}\hat{\sigma}_z + \hbar\omega_1 \cos(\omega_{RF} t)\hat{\sigma}_x, \tag{9.12}$$

$$= \frac{\hbar\Delta\omega}{2}\hat{\sigma}_z + \frac{\hbar\omega_{RF}}{2}\hat{\sigma}_z + \hbar\omega_1 \cos(\omega_{RF} t)\hat{\sigma}_x, \tag{9.13}$$

where ω_0 and ω_1 are the Larmor frequencies associated with the fields $\boldsymbol{B}_0$ and $\boldsymbol{B}_1(\omega_{RF})$, respectively, as defined in Eq. (9.5), and $\Delta\omega = \omega_0 - \omega_{RF}$. The typical values of the involved fields are $B_0 \sim 0.1 \div 10$ T and $B_1 \sim 10\,\mu$T $\div$ 100 mT, thus, $B_0 \gg B_1$.

Now, we consider the interaction picture with respect to $\hat{H} = (\hbar\omega_{RF}/2)\hat{\sigma}_z$ and the Hamiltonian $\hat{H}_{NMR}$ transforms into (see Appendix A for details):

$$\hat{H}'_{NMR} = \frac{\hbar\Delta\omega}{2}\hat{\sigma}_z + \frac{\hbar\omega_1}{2}\left\{[1 + \cos(2\omega_{RF} t)]\hat{\sigma}_x - \sin(2\omega_{RF} t)\hat{\sigma}_y\right\}, \tag{9.14}$$

which, performing the *rotating-wave approximation* (RWA) or *secular approximation*, i.e., neglecting the fast rotating terms involving $2\omega_{RF} t$, reduces to:

$$\hat{H}'_{NMR} = \frac{\hbar\Delta\omega}{2}\hat{\sigma}_z + \frac{\hbar\omega_1}{2}\hat{\sigma}_x, \tag{9.15}$$

or, simply:

$$\hat{H}'_{NMR} = \frac{\hbar\omega_{eff}}{2}\boldsymbol{n}\cdot\hat{\boldsymbol{\sigma}}, \tag{9.16}$$

where we introduced the effective frequency:

$$\omega_{eff} = \sqrt{\omega_1^2 + (\Delta\omega)^2}, \tag{9.17}$$

and:

$$\boldsymbol{n} = \frac{1}{\omega_{eff}}(\omega_1, 0, \Delta\omega). \tag{9.18}$$

We can conclude that we have a precession of the spin around the direction given by $\boldsymbol{n}$. By suitably changing $\Delta\omega$ it is possible to vary $\boldsymbol{n}$ and, thus, to implement different single-qubit operations.

9.1.4 How to Realize a CNOT Gate with Spin Systems

The CNOT gate involves two qubits and the corresponding operator, taking qubits 1 and 2 as control and target, respectively, may be written as the following operator:

$$\mathbf{C}_{12} = \frac{1}{2}\left(\hat{\mathbb{I}} + \hat{\sigma}_z^{(1)} + \hat{\sigma}_x^{(2)} - \hat{\sigma}_z^{(1)}\hat{\sigma}_x^{(2)}\right), \tag{9.19}$$

where $\hat{\sigma}_k^{(h)}$, $k = x, y, z$ and $h = 1, 2$, represent the Pauli operators acting on the h-th qubit (see Sect. 1.4.2). However, as mentioned in Sect. 3.5, $\hat{\sigma}_z = \mathbf{H}\hat{\sigma}_x\mathbf{H}$, therefore we can focus on the operator:

$$\mathbf{Z}_{12} = \left(\hat{\mathbb{I}} \otimes \mathbf{H}\right)\mathbf{C}_{12}\left(\hat{\mathbb{I}} \otimes \mathbf{H}\right), \tag{9.20a}$$

$$= \frac{1}{2}\left(\hat{\mathbb{I}} + \hat{\sigma}_z^{(1)} + \hat{\sigma}_z^{(2)} - \hat{\sigma}_z^{(1)}\hat{\sigma}_z^{(2)}\right), \tag{9.20b}$$

which is symmetric with respect the exchange of the two qubits. Since $(\mathbf{Z}_{12})^2 = \hat{\mathbb{I}}$ we have:

$$\exp(i\,\mathbf{Z}_{12}\,\theta) = \sum_{k=0}^{\infty}\frac{(i\theta)^k}{k!}(\mathbf{Z}_{12})^k \tag{9.21a}$$

$$= \cos\theta\,\hat{\mathbb{I}} + i\,\mathbf{Z}_{12}\,\sin\theta, \tag{9.21b}$$

and, setting $\theta = \pi/2$, we find:

$$\mathbf{Z}_{12} = -i\exp\left(i\,\mathbf{Z}_{12}\,\frac{\pi}{2}\right) \tag{9.22a}$$

$$= -i\exp\left[i\frac{\pi}{4}\left(\hat{\mathbb{I}} + \hat{\sigma}_z^{(1)} + \hat{\sigma}_z^{(2)} - \hat{\sigma}_z^{(1)}\hat{\sigma}_z^{(2)}\right)\right] \tag{9.22b}$$

$$= \exp\left(-i\frac{\pi}{4}\right)\exp\left[i\frac{\pi}{4}\left(\hat{\sigma}_z^{(1)} + \hat{\sigma}_z^{(2)} - \hat{\sigma}_z^{(1)}\hat{\sigma}_z^{(2)}\right)\right]. \tag{9.22c}$$

Therefore, we can implement the $\mathbf{Z}_{12}$ gate by letting the two qubits interact through the Hamiltonian:

$$\hat{H} \propto \hat{\sigma}_z^{(1)} + \hat{\sigma}_z^{(2)} - \hat{\sigma}_z^{(1)}\hat{\sigma}_z^{(2)}, \tag{9.23}$$

and by choosing a suitable time t for the corresponding unitary evolution. As we will see, the term $\hat{H}_0 \propto \hat{\sigma}_z^{(1)} + \hat{\sigma}_z^{(2)}$ is the free Hamiltonian of the system of the two qubits, while $\hat{\sigma}_z^{(1)}\hat{\sigma}_z^{(2)}$ represents a highly anisotropic interaction that couples the z-components of the qubits, known as *Ising interaction*.

Physically, the Hamiltonian $\hat{H}$ may be realized with spin-1/2 particles. In this case the free Hamiltonian is

$$\hat{H}_0 \propto \hat{\sigma}_z^{(1)} + \hat{\sigma}_z^{(2)}, \tag{9.24}$$

and the interaction Hamiltonian $\hat{\sigma}_z^{(1)}\hat{\sigma}_z^{(2)}$ couples the spins along the z-direction subject to an uniform magnetic field, whose amplitude is proportional to the strength of their coupling. However, Ising interactions are hard to arrange and it is better to consider *exchange interactions* between spins. As we will see in Chap. 9 (Sect. 9.1.5), by applying suitable magnetic fields to the spins, with the same direction but different magnitudes and signs, we can build a $\mathbf{Z}_{12}$ gate.

9.1.5 Exchange Interactions and CNOT Gate

In Sect. 9.1.4 we have seen that CNOT may be implemented with two spin–1/2 systems by using the Ising interaction, that is a kind of interaction which couples spin along z-direction. However, we pointed out that Ising interactions are hard to arrange and it is better to use *exchange interactions* between two spins, whose interaction Hamiltonian is:

$$\hat{H}_{\text{ex}} \propto \hat{\boldsymbol{\sigma}}^{(1)} \cdot \hat{\boldsymbol{\sigma}}^{(2)} = \hat{\sigma}_x^{(1)}\hat{\sigma}_x^{(2)} + \hat{\sigma}_y^{(1)}\hat{\sigma}_y^{(2)} + \hat{\sigma}_z^{(1)}\hat{\sigma}_z^{(2)}, \tag{9.25}$$

where

$$\hat{\boldsymbol{\sigma}}^{(k)} = \left(\hat{\sigma}_x^{(k)}, \hat{\sigma}_y^{(k)}, \hat{\sigma}_z^{(k)} \right), \tag{9.26}$$

with $k = 1, 2$, is the vector of the Pauli operators acting on the Hilbert space $\mathcal{H}_k$ of the k-th spin.

The system we are considering here consists of two spin–1/2 particles of mass m_k and charge q_k, $k = 1, 2$. We assume that each spin (resonantly) interacts with a magnetic field $\boldsymbol{B}_k$ whereas they are coupled through exchange interaction. The corresponding Hamiltonian reads (we use the same formalism introduced in the previous sections):

$$\hat{H} = \frac{\hbar\omega_1}{2}\boldsymbol{n}_1 \cdot \hat{\boldsymbol{\sigma}}^{(1)} + \frac{\hbar\omega_2}{2}\boldsymbol{n}_2 \cdot \hat{\boldsymbol{\sigma}}^{(2)} + \hbar J\, \hat{\boldsymbol{\sigma}}^{(1)} \cdot \hat{\boldsymbol{\sigma}}^{(2)}, \tag{9.27}$$

where ω_k are the corresponding Larmor frequencies and J is the strength of the exchange interaction. Note that if $J = 0$, then Eq. (9.27) reduces to the Hamiltonian of two uncoupled spins each interacting with the corresponding magnetic field and we have just two independent single-qubit gates.

9.1 Universal Computation with Spins

Without lack of generality we can set $\boldsymbol{B}_k = (0, 0, B_k)$ and Eq. (9.27) becomes:

$$\hat{H} = \underbrace{\frac{\hbar\omega_1}{2}\,\hat{\sigma}_z^{(1)} + \frac{\hbar\omega_2}{2}\,\hat{\sigma}_z^{(2)}}_{\hat{H}_0} + \underbrace{\hbar J\,\hat{\boldsymbol{\sigma}}^{(1)} \cdot \hat{\boldsymbol{\sigma}}^{(2)}}_{\hat{H}_{\text{ex}}}, \tag{9.28}$$

where $\hat{H}_0$ is the free Hamiltonian of the two-spin system, while $\hat{H}_{\text{ex}}$ is the interaction Hamiltonian. In the following we show that, starting form the Hamiltonian in Eq. (9.28), we can build the two-qubit quantum gate $\mathbf{Z}_{12}$, that can be converted into a CNOT gate by means of Hadamard transformations realized through Eq. (9.11) (see Sect. 9.1.4). In particular, we show that for a suitable choice of ω_k and t, given J, we may have $\mathbf{Z}_{12} = \exp(-i\,\hat{H}t/\hbar)$. First of all, we recall that:

$$\mathbf{Z}_{12} = \frac{1}{2}\left(\hat{\mathbb{I}} + \hat{\sigma}_z^{(1)} + \hat{\sigma}_z^{(2)} - \hat{\sigma}_z^{(1)}\hat{\sigma}_z^{(2)}\right), \tag{9.29}$$

and we have:

$$\mathbf{Z}_{12}|00\rangle = |00\rangle, \quad \mathbf{Z}_{12}|11\rangle = -|11\rangle, \tag{9.30a}$$

$$\mathbf{Z}_{12}|\psi_+\rangle = |\psi_+\rangle, \quad \mathbf{Z}_{12}|\psi_-\rangle = |\psi_-\rangle, \tag{9.30b}$$

where:

$$|00\rangle, \quad |11\rangle \quad \text{and} \quad |\psi_+\rangle = \frac{|01\rangle + |10\rangle}{\sqrt{2}} \tag{9.31}$$

are the *triplet states* and:

$$|\psi_-\rangle = \frac{|01\rangle - |10\rangle}{\sqrt{2}} \tag{9.32}$$

is the *singlet state*. It is worth noting that the four states $\{|00\rangle, |11\rangle, |\psi_\pm\rangle\}$ form a basis of the Hilbert space $\mathcal{H}_1 \otimes \mathcal{H}_2$, $\mathcal{H}_k$ being the Hilbert space of the k-th spin. Therefore, it is enough to find the conditions on the involved parameters in order to have the evolution operator

$$U_{\text{ex}}(t) = \exp\left(-\frac{i}{\hbar}\hat{H}t\right) \tag{9.33}$$

acting as $\mathbf{Z}_{12}$ on such a basis.

The first step is to find the eigenvectors and eigenvalues of Eq. (9.28) and we proceed as follows. Since the SWAP operator may be written as:

$$\mathbf{S} = \frac{1}{2}\left(\hat{\mathbb{I}} + \hat{\boldsymbol{\sigma}}^{(1)} \cdot \hat{\boldsymbol{\sigma}}^{(2)}\right), \tag{9.34}$$

the following states are eigenstates of the operator $\hat{\boldsymbol{\sigma}}^{(1)} \cdot \hat{\boldsymbol{\sigma}}^{(2)}$, namely:

$$\hat{\boldsymbol{\sigma}}^{(1)} \cdot \hat{\boldsymbol{\sigma}}^{(2)} |00\rangle = |00\rangle, \qquad \hat{\boldsymbol{\sigma}}^{(1)} \cdot \hat{\boldsymbol{\sigma}}^{(2)} |11\rangle = |11\rangle, \tag{9.35a}$$

$$\hat{\boldsymbol{\sigma}}^{(1)} \cdot \hat{\boldsymbol{\sigma}}^{(2)} |\psi_+\rangle = |\psi_+\rangle, \qquad \hat{\boldsymbol{\sigma}}^{(1)} \cdot \hat{\boldsymbol{\sigma}}^{(2)} |\psi_-\rangle = -3|\psi_-\rangle. \tag{9.35b}$$

Furthermore, we can write:

$$\hat{H}_0 = \frac{\hbar\omega_+}{2}\left(\frac{\hat{\sigma}_z^{(1)} + \hat{\sigma}_z^{(2)}}{2}\right) + \frac{\hbar\omega_-}{2}\left(\frac{\hat{\sigma}_z^{(1)} - \hat{\sigma}_z^{(2)}}{2}\right), \tag{9.36}$$

with $\omega_\pm = \omega_1 \pm \omega_2$ and we find:

$$\frac{1}{2}\left(\hat{\sigma}_z^{(1)} + \hat{\sigma}_z^{(2)}\right)|00\rangle = |00\rangle, \qquad \frac{1}{2}\left(\hat{\sigma}_z^{(1)} - \hat{\sigma}_z^{(2)}\right)|00\rangle = 0, \tag{9.37a}$$

$$\frac{1}{2}\left(\hat{\sigma}_z^{(1)} + \hat{\sigma}_z^{(2)}\right)|11\rangle = -|11\rangle, \qquad \frac{1}{2}\left(\hat{\sigma}_z^{(1)} - \hat{\sigma}_z^{(2)}\right)|11\rangle = 0, \tag{9.37b}$$

$$\frac{1}{2}\left(\hat{\sigma}_z^{(1)} + \hat{\sigma}_z^{(2)}\right)|\psi_\pm\rangle = 0, \qquad \frac{1}{2}\left(\hat{\sigma}_z^{(1)} - \hat{\sigma}_z^{(2)}\right)|\psi_\pm\rangle = |\psi_\mp\rangle. \tag{9.37c}$$

Therefore we have:

$$\hat{H}|00\rangle = \hbar\left(J + \frac{\omega_+}{2}\right)|00\rangle, \qquad \hat{H}|11\rangle = \hbar\left(J - \frac{\omega_+}{2}\right)|11\rangle, \tag{9.38a}$$

$$\hat{H}|\psi_+\rangle = \hbar J|\psi_+\rangle + \frac{\hbar\omega_-}{2}|\psi_-\rangle, \qquad \hat{H}|\psi_-\rangle = -3\hbar J|\psi_-\rangle + \frac{\hbar\omega_-}{2}|\psi_+\rangle, \tag{9.38b}$$

that is $|00\rangle$ and $|11\rangle$ are eigenstates of $\hat{H}$, while $\hat{H}$ transforms $|\psi_\pm\rangle$ is a linear combination of $|\psi_+\rangle$ and $|\psi_-\rangle$.

Thereafter, we have the following matrix representation of $\hat{H}$ in the chosen basis:

$$\hat{H} \to \hbar \begin{pmatrix} J + \frac{1}{2}\omega_+ & 0 & 0 & 0 \\ 0 & J - \frac{1}{2}\omega_+ & 0 & 0 \\ 0 & 0 & J & \frac{1}{2}\omega_- \\ 0 & 0 & \frac{1}{2}\omega_- & -3J \end{pmatrix}. \tag{9.39}$$

9.1 Universal Computation with Spins

The matrix has a block-diagonal form and, to find its eigenvectors and eigenvalues, we can consider only the 2×2 block [the other block is with eigenvectors and eigenvalues given in Eq. (9.38a)]:

$$\begin{pmatrix} J & \frac{1}{2}\omega_- \\ \frac{1}{2}\omega_- & -3J \end{pmatrix} \quad (9.40)$$

that has eigenvalues:

$$-J \pm \sqrt{4J^2 + \frac{1}{4}\omega_-^2}, \quad (9.41)$$

corresponding to the eigenstates:

$$|\Psi_\pm\rangle = \alpha_\pm |\psi_+\rangle + \beta_\pm |\psi_-\rangle, \quad (9.42)$$

where we do not explicitly calculate the expression of the coefficients $\alpha_\pm$ and $\beta_\pm$. Now, since $|\psi_\pm\rangle$ are eigenstates of $\mathbf{Z}_{12}$ with eigenvalue 1 [see Eq. (9.30b)], the states $|\Psi_\pm\rangle$ are still its eigenstates with the same eigenvalue. Therefore, we have found that the four states:

$$|00\rangle, \quad |11\rangle, \quad \text{and} \quad |\Psi_\pm\rangle, \quad (9.43)$$

are eigenstates of both $\mathbf{Z}_{12}$ and $\hat{H}$ and, thus, of the evolution operator $U_{\text{ex}}(t)$.

In order to have $\mathbf{Z}_{12} \equiv U_{\text{ex}}(t)$, their eigenstates should have the same eigenvalues, up to a constant phase factor which should be the same for all the states, namely:

$$U_{\text{ex}}(t)|00\rangle = \exp\left[-it\left(J + \frac{1}{2}\omega_+\right)\right]|00\rangle \quad \leftrightarrow \quad \mathbf{Z}_{12}|00\rangle = |00\rangle, \quad (9.44a)$$

$$U_{\text{ex}}(t)|11\rangle = \exp\left[-it\left(J - \frac{1}{2}\omega_+\right)\right]|11\rangle \quad \leftrightarrow \quad \mathbf{Z}_{12}|11\rangle = -|11\rangle, \quad (9.44b)$$

$$U_{\text{ex}}(t)|\Psi_+\rangle = \exp\left[-it\left(-J + \sqrt{4J^2 + \frac{1}{4}\omega_-^2}\right)\right]|\Psi_+\rangle \quad \leftrightarrow \quad \mathbf{Z}_{12}|\Psi_+\rangle = |\Psi_+\rangle, \quad (9.44c)$$

$$U_{\text{ex}}(t)|\Psi_-\rangle = \exp\left[-it\left(-J - \sqrt{4J^2 + \frac{1}{4}\omega_-^2}\right)\right]|\Psi_-\rangle \quad \leftrightarrow \quad \mathbf{Z}_{12}|\Psi_-\rangle = |\Psi_-\rangle. \quad (9.44d)$$

This happens by setting

$$\omega_+ = 4J, \quad \omega_- = 4\sqrt{3}J, \quad \text{and} \quad t = \frac{\pi}{4J}, \quad (9.45)$$

which also leads to the overall constant phase factor $\exp(-i\,3\pi/4)$ equal for all the states. Indeed, one can change the value of $\omega_\pm$ by varying the amplitudes of the two magnetic fields. In fact, the previous conditions are equivalent to require:

$$\omega_1 = 2\left(1+\sqrt{3}\right)J \quad \text{and} \quad \omega_2 = 2\left(1-\sqrt{3}\right)J, \quad (9.46)$$

and, thus, we find:

$$B_1 = 4\left(\sqrt{3}+1\right)\frac{mJ}{gq}, \quad \text{and} \quad B_2 = -4\left(\sqrt{3}-1\right)\frac{mJ}{gq}, \quad (9.47)$$

where, for the sake of simplicity, we assumed the two spin–1/2 particles to be of the same species, i.e., $m_k = m$, $g_k = g$ and $q_k = q$, $k = 1, 2$. Note that the two magnetic fields are directed along z-direction but have opposite sign; though $\mathbf{Z}_{12}$ is symmetric, its physical implementation by means of exchange interaction requires different magnetic fields acting on the two spins. However, if we set:

$$\omega_1 = 2\left(1-\sqrt{3}\right)J \quad \text{and} \quad \omega_2 = 2\left(1+\sqrt{3}\right)J, \quad (9.48)$$

we obtain the same result, that is, the symmetry is still present!

Let us now focus on the order of magnitude of the involved quantities. The Bohr magneton and the nuclear magneton are:

$$\mu_B = \frac{e\hbar}{2m_e} = 9.27 \times 10^{-24} \frac{J}{T} \quad \text{and} \quad \mu_N = \frac{e\hbar}{2m_p} = 5.05 \times 10^{-27} \frac{J}{T} \quad (9.49)$$

respectively, where e is the charge of the electron while m_e and m_p are the masses of the electron and of the proton, respectively. Typical spin–1/2 nuclei are ^{1}H, ^{13}C and ^{19}F and the J-coupling magnitudes are $J \sim 10^8$ Hz ($\sim$100 MHz). Since $\omega \sim 10^8$ Hz, we have that the involved magnetic field amplitudes are $\sim 10^{-2}$ T for the electronic spin and ~ 10 T for the nuclear spin, leading to a time-scale $t \sim 10^{-8}$ sec.

9.1.6 Further Considerations

The exchange interaction Hamiltonians are typical of NMR systems and molecules. The interaction between the spins is an indirect interaction mediated by the electrons shared through a chemical bond. The magnetic field seen by the nucleus is perturbed by the state of the electronic cloud, which interacts with another nucleus through the overlap of the wave-function with the nucleus (Fermi contact interaction), that is a through-bond interaction.

The same Hamiltonian of Eq. (9.27) describes the excess of electron spins in pair of quantum dots, which are linked through a tunnel junction (Heisenberg Hamiltonian). This effective Hamiltonian can be derived from a microscopic model for electrons in coupled quantum dots.

9.2 Interaction Between Atoms and Light: Cavity QED

Here we address a two-level atom as a qubit. Throughout this section $|g\rangle$ and $|e\rangle$ represent the states associated with the ground and the excited state, respectively. The free Hamiltonian of the two-level atom can be written by means of the Pauli operators as follows:

$$\hat{H}_a = \frac{\hbar \omega_{eg}}{2} \hat{\sigma}_z, \qquad (9.50)$$

where $\hbar \omega_{eg} = \hbar \omega_e - \hbar \omega_g$ is the energy difference between the two levels and we have the following association with the usual computational basis:

$$|e\rangle \to |0\rangle \quad \text{and} \quad |g\rangle \to |1\rangle. \qquad (9.51)$$

In the two-level approximation, the electric-dipole moment operator of the atom can be written as:

$$\hat{\boldsymbol{D}} = d\,(\boldsymbol{\varepsilon}_a \hat{\sigma}_- + \boldsymbol{\varepsilon}_a^* \hat{\sigma}_+), \qquad (9.52)$$

where we introduced:

$$\hat{\sigma}_- = |g\rangle\langle e| \quad \text{and} \quad \hat{\sigma}_+ = |e\rangle\langle g|, \qquad (9.53)$$

the lowering and raising operators, d is the matrix element of the atomic transition and $\boldsymbol{\varepsilon}_a$ is a complex vector which represents the atomic polarization transition. Note that:

$$\hat{\sigma}_\pm = \frac{\hat{\sigma}_x \pm i \hat{\sigma}_y}{2}. \qquad (9.54)$$

9.2.1 Interaction Between a Two-Level Atom and a Classical Electric Field

The interaction between a two-level atom and a classical electric field is formally equivalent to the interaction between a spin–1/2 particle and a magnetic field discussed in the previous sections. The quantum Hamiltonian describing the interaction between the atomic electric dipole moment and the classical field $\boldsymbol{E}(t)$ inside a cavity (see Appendix B) is:

$$\hat{H}_{\text{int}} = -\hat{\boldsymbol{D}} \cdot \boldsymbol{E}(t), \tag{9.55}$$

with:

$$\boldsymbol{E}(t) = i \, E_0 \left(\boldsymbol{\varepsilon}_{\text{f}} \, e^{-i\omega t - i\varphi} - \boldsymbol{\varepsilon}_{\text{f}}^* \, e^{i\omega t + i\varphi} \right), \tag{9.56}$$

where ω and $\boldsymbol{\varepsilon}_{\text{f}}$ are the frequency and the polarization of the field, respectively, and we assumed a real amplitude E_0. The whole Hamiltonian of the system is thus given by:

$$\hat{H}_{\text{tot}} = \frac{\hbar \omega_{eg}}{2} \hat{\sigma}_z - \hat{\boldsymbol{D}} \cdot \boldsymbol{E}(t), \tag{9.57a}$$

$$= \frac{\hbar \Delta \omega}{2} \hat{\sigma}_z + \frac{\hbar \omega}{2} \hat{\sigma}_z - \hat{\boldsymbol{D}} \cdot \boldsymbol{E}(t), \tag{9.57b}$$

where $\Delta \omega = \omega_{eg} - \omega$ is the detuning between the two-level atom and the field. In order to focus on the interaction, we consider the interaction picture with respect to the Hamiltonian $\hat{H}_0 = \hbar \omega \hat{\sigma}_z / 2$: it is worth noting that here we are considering the frequency ω of the field. Following Appendix A and passing to the interaction picture, we have:

$$\hat{H}_{\text{tot}} \to \hat{H} = \hat{U}_0^\dagger(t) \, \hat{H}_{\text{tot}} \, \hat{U}_0(t) = \frac{\hbar \Delta \omega}{2} \hat{\sigma}_z - \hat{U}_0^\dagger(t) \, \hat{\boldsymbol{D}} \cdot \boldsymbol{E}(t) \, \hat{U}_0(t). \tag{9.58}$$

Since

$$\hat{U}_0^\dagger(t) \, \hat{\sigma}_\pm \, \hat{U}_0(t) = \hat{\sigma}_\pm \, e^{\pm i\omega t}, \tag{9.59}$$

the last term of Eq. (9.58) contains terms proportional to $e^{\pm i\varphi}$ and to $e^{\pm i 2\omega t \pm i\varphi}$: these last terms are *fast rotating* and if we assume that the time-scale of the system is $1/\omega$, then their effect on the time evolution is negligible. This corresponds to perform the RWA and Eq. (9.57) reduces to (for the sake of simplicity we assume $\boldsymbol{\varepsilon}_{\text{a}}, \boldsymbol{\varepsilon}_{\text{f}} \in \mathbb{R}^3$):

$$\hat{H} = \frac{\hbar \Omega'}{2} \boldsymbol{n} \cdot \hat{\boldsymbol{\sigma}}, \tag{9.60}$$

where:

$$\boldsymbol{n} = \frac{1}{\Omega'} \left(-\Omega_0 \sin \varphi, \, \Omega_0 \cos \varphi, \, \Delta \omega \right). \tag{9.61}$$

9.2 Interaction Between Atoms and Light: Cavity QED

with

$$\Omega' = \sqrt{(\Delta\omega)^2 + \Omega_0^2} \qquad (9.62)$$

and we introduced the *Rabi frequency*:

$$\Omega_0 = \frac{2d}{\hbar} E_0 \, \boldsymbol{\varepsilon}_a \cdot \boldsymbol{\varepsilon}_f. \qquad (9.63)$$

In the resonant case ($\Delta\omega = 0$) we have (we can set $\varphi = 0$):

$$\hat{H} = \frac{\hbar\Omega_0}{2} \hat{\sigma}_y, \qquad (9.64)$$

which clearly has the following eigenstates:

$$|\gamma_\pm\rangle = \frac{1}{\sqrt{2}}(|0\rangle \pm i|1\rangle). \qquad (9.65)$$

More in general, if $\varphi \neq 0$, we obtain the following time evolution (still in the resonant case):

$$\hat{U}_\varphi(t) = \exp\left(-i\frac{\Omega_0 t}{2} \boldsymbol{n} \cdot \hat{\boldsymbol{\sigma}}\right),$$

$$= \cos\left(\frac{\Omega_0 t}{2}\right) \hat{\mathbb{1}} - i \sin\left(\frac{\Omega_0 t}{2}\right) \left(-\sin\varphi \, \hat{\sigma}_x + \cos\varphi \, \hat{\sigma}_y\right), \qquad (9.66)$$

and, by using the 2×2 matrix formalism (in the computational basis):

$$\hat{U}_\varphi(t) \to \begin{pmatrix} \cos\left(\frac{\Omega_0 t}{2}\right) & -e^{-i\varphi} \sin\left(\frac{\Omega_0 t}{2}\right) \\ e^{i\varphi} \sin\left(\frac{\Omega_0 t}{2}\right) & \cos\left(\frac{\Omega_0 t}{2}\right) \end{pmatrix}. \qquad (9.67)$$

It is now straightforward to see that, in the interaction picture:

$$\hat{U}_\varphi(t)|e\rangle = \cos\left(\frac{\Omega_0 t}{2}\right) |e\rangle + e^{i\varphi} \sin\left(\frac{\Omega_0 t}{2}\right) |g\rangle, \qquad (9.68a)$$

$$\hat{U}_\varphi(t)|g\rangle = \cos\left(\frac{\Omega_0 t}{2}\right) |g\rangle - e^{-i\varphi} \sin\left(\frac{\Omega_0 t}{2}\right) |e\rangle. \qquad (9.68b)$$

We have three following relevant cases.

- $\frac{\pi}{2}$-pulse: in this case one sets $\Omega_0 t = \pi/2$ and we have the following evolution starting from $|g\rangle$ or $|e\rangle$:

$$|e\rangle \to \frac{1}{\sqrt{2}}\left(|e\rangle + e^{i\varphi}|g\rangle\right), \quad \text{and} \quad |g\rangle \to \frac{1}{\sqrt{2}}\left(|g\rangle - e^{-i\varphi}|e\rangle\right), \quad (9.69)$$

and, for $\varphi = 0$, we obtain the Hadamard transformation.

- π-pulse: now $\Omega_0 t = \pi$ and we have:

$$|e\rangle \to e^{i\varphi}|g\rangle, \quad \text{and} \quad |g\rangle \to -e^{-i\varphi}|e\rangle, \quad (9.70)$$

that is, besides and overall phase shift, the NOT gate.

- 2π-pulse: for $\Omega_0 t = 2\pi$ we get:

$$|e\rangle \to -|e\rangle, \quad \text{and} \quad |g\rangle \to -|g\rangle, \quad (9.71)$$

i.e., we add a phase shift to the input state. This phase shift is a well-known properties of 2π-spin rotations.

9.3 The Quantum Description of Light

The quantum Hamiltonian of the single-mode electromagnetic field corresponds to that of a harmonic oscillator with the same frequency ω, namely:

$$\hat{H} = \frac{\hat{P}^2}{2} + \frac{1}{2}\omega^2 \hat{Q}^2 = \hbar\omega\left(\hat{a}^\dagger \hat{a} + \frac{1}{2}\right) \quad (9.72)$$

where we introduced the position- and momentum-like operators:

$$\hat{Q} = \sqrt{\frac{\hbar}{2\omega}}\left(\hat{a}^\dagger + \hat{a}\right) \quad \text{and} \quad \hat{P} = i\sqrt{\frac{\hbar\omega}{2}}\left(\hat{a}^\dagger - \hat{a}\right), \quad (9.73)$$

respectively, $\left[\hat{Q}, \hat{P}\right] = i\hbar \hat{\mathbb{I}}$, and:

$$\hat{a} = \sqrt{\frac{\omega}{2\hbar}}\left(\hat{Q} + i\frac{\hat{P}}{\omega}\right) \quad \text{and} \quad \hat{a}^\dagger = \sqrt{\frac{\omega}{2\hbar}}\left(\hat{Q} - i\frac{\hat{P}}{\omega}\right), \quad (9.74)$$

are the annihilation and creation bosonic field operators respectively, $[\hat{a}, \hat{a}^\dagger] = \hat{\mathbb{I}}$. More in general, to each mode of the radiation field corresponds a bosonic field operator.

9.4 Photonic Qubits

If we denote with $\{|n\rangle\}_{n\in\mathbb{N}}$ the set of the eigenvectors of the self-adjoint operator $\hat{N} = \hat{a}^\dagger \hat{a}$, namely, $\hat{N}|n\rangle = n|n\rangle$, we have:

$$\hat{a}|n\rangle = \sqrt{n}|n-1\rangle \quad \text{and} \quad \hat{a}^\dagger|n\rangle = \sqrt{n+1}|n+1\rangle, \tag{9.75}$$

and, thus:

$$|n\rangle = \frac{(\hat{a}^\dagger)^n}{\sqrt{n!}}|0\rangle, \tag{9.76}$$

where the state $|0\rangle$ represents the vacuum state. The set $\{|n\rangle\}_{n\in\mathbb{N}}$ is sometimes called Fock-state basis or photon-number basis.

Beside the frequency, the photons have also a polarization degree of freedom, that can be used to encode qubits. Given a single-photon state, we associate the polarization with a two dimensional Hilbert space spanned by the basis vectors $\{|H\rangle, |V\rangle\}$ and we have a direct correspondence with the qubit states $\{|0\rangle, |1\rangle\}$: this is a polarization qubit.

Usually, when it is clear that we are dealing with a single photon, one can represent its state as a vector in the polarization Hilbert state, without specifying the actual photon number. Thereafter, the single-photon superposition state:

$$|\psi\rangle = \cos\theta\,|H\rangle + \sin\theta\,|V\rangle, \tag{9.77}$$

physically represents a photon with linear polarization at an angle θ with respect to the horizontal one. A quantum superposition with complex coefficients, such as:

$$|\psi\rangle = \cos\theta\,|H\rangle + i\,\sin\theta\,|V\rangle, \tag{9.78}$$

corresponds to an elliptic polarization, that reduces to the circular one if $\theta = \pi/4$. Moreover, single-photon optical qubits can exploit the orbital angular momentum (of light) degree of freedom and their direction of propagation to encode information (path qubits).

Emanuel Knill, Raymond Laflamme and Gerard J. Milburn have shown that universal quantum computation can be implemented exploiting linear optical tools, such as beam spitters and phase shifters, photon detectors and, of course, single photon sources. This is known as KLM protocol and it is a particular case of linear optical quantum computing.

Still in the realm of linear optics, a limited model of non-universal quantum computation has been proposed by Scott Aaronson and Alex Arkhipov: the boson sampling. The so-called boson sampling problem is a problem believed to be beyond the ability of classical computers. Its implementations are not only limited to single photons but can exploit the class of the Gaussian states (and photon detection), namely, states exhibiting Gaussian Wigner functions, such as squeezed states. We recall that the coherent states of light, a good approximation of the states generated by a laser, are Gaussian states.

Before closing this section, we mention the Gottesman–Kitaev–Preskill (GKP) code, proposed by Daniel Gottesman, Alexei Kitaev and John Preskill that is an interesting way to implement a practical, fault-tolerant quantum computation (see Sect. 8.4). The code exploits the state of an oscillator and the GKP codewords are coherent superpositions of periodically displaced squeezed vacuum states.

9.5 The Jaynes–Cummings Model

The full quantum model to describe the interaction between light and matter involves the quantum description of light. Now the classical electric field appearing in the interaction Hamiltonian of Eq. (9.55) is replaced by the corresponding quantum operator:[1]

$$\hat{E} = i E_0 \left(\boldsymbol{\varepsilon}_f \hat{a} - \boldsymbol{\varepsilon}_f^* \hat{a}^\dagger \right), \tag{9.79}$$

where $\hat{a}$ and $\hat{a}^\dagger$ are the annihilation and creation field operators introduced in Sect. 9.3 describing the stationary field inside the cavity (we assume the atom at the cavity center). The free Hamiltonian of the system reads:

$$\hat{H}_0 = \underbrace{\frac{\hbar \omega_{eg}}{2} \hat{\sigma}_z}_{\text{atom}} + \underbrace{\hbar \omega \left(\hat{a}^\dagger \hat{a} + \frac{1}{2} \right)}_{\text{field}}, \tag{9.80}$$

and we have two families of eigenstates of $\hat{H}_0$, i.e.:

$$\hat{H}_0 |g, n\rangle = \hbar \left[-\frac{\omega_{eg}}{2} + \omega \left(n + \frac{1}{2} \right) \right] |g, n\rangle, \tag{9.81a}$$

$$\hat{H}_0 |e, n\rangle = \hbar \left[+\frac{\omega_{eg}}{2} + \omega \left(n + \frac{1}{2} \right) \right] |e, n\rangle, \tag{9.81b}$$

where $\{|e\rangle, |g\rangle\}$ are the eigenstates of $\hat{\sigma}_z$, $\{|n\rangle\}$ is the photon-number basis and $|x, y\rangle = |x\rangle|y\rangle$. As we can also see in Fig. 9.2, if $\omega = \omega_{eg}$ the states $|g, n+1\rangle$ and $|e, n\rangle$, with $n \geq 0$, are degenerate.

The interaction Hamiltonian reads:

$$H_{\text{int}} = -\hat{\boldsymbol{D}} \cdot \hat{\boldsymbol{E}}, \tag{9.82}$$

---

[1] We consider a stationary, time-independent cavity field and, for the sake of simplicity, we also assume that the atom is placed at the center of the cavity.

9.5 The Jaynes–Cummings Model

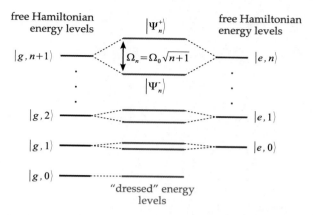

Fig. 9.2 The blue and red lines refer to the energy levels corresponding to the eigenstates of the free Hamiltonian given in Eq. (9.80) with $\omega = \omega_{eg}$: it is clear that the states $|g, n+1\rangle$ and $|e, n\rangle$, with $n \geq 0$, are degenerate. The only non-degenerate level is the ground state $|g, 0\rangle$. The Jaynes–Cummings interaction removes degeneracy and couples the dressed states $|\Psi_n^{\pm}\rangle$, whose corresponding energy levels (green lines) have an energy difference equal to $\hbar\Omega_n = \hbar\Omega_0\sqrt{n+1}$

where $\hat{D}$ is still given by Eq. (9.52). By performing the interaction picture with respect to the Hamiltonian (see Appendix A and Sect. 9.2.1):

$$\hat{H}' = \hbar\omega \left(\hat{a}^{\dagger}\hat{a} + \frac{1}{2}\hat{\mathbb{I}} + \frac{1}{2}\hat{\sigma}_z \right), \tag{9.83}$$

and the RWA, we obtain the following interaction Hamiltonian:

$$\hat{H}_{\text{int}} = \frac{\hbar\delta}{2}\hat{\sigma}_z - i\frac{\hbar\Omega_0}{2} \left(\hat{\sigma}_+\hat{a} - \hat{\sigma}_-\hat{a}^{\dagger} \right), \tag{9.84}$$

(Jaynes–Cummings Hamiltonian)

where Ω_0 is the Rabi frequency defined in Eq. (9.63) and $\delta = \omega_{eg} - \omega$ is the *detuning*.

It is interesting to note that $\hat{H}_{\text{int}}$ couples the two-dimensional manifold spanned by $\{|g, n+1\rangle, |e, n\rangle\}$, with $n > 0$. In fact, we have:

$$\left(\hat{\sigma}_+\hat{a} - \hat{\sigma}_-\hat{a}^{\dagger} \right) |g, n+1\rangle = \sqrt{n+1}\,|e, n\rangle,$$

(absorption of one photon) (9.85)

and

$$\left(\hat{\sigma}_+\hat{a} - \hat{\sigma}_-\hat{a}^{\dagger} \right) |e, n\rangle = -\sqrt{n+1}\,|g, n+1\rangle.$$

(emission of one photon) (9.86)

Note that the ground state of the free Hamiltonian, namely, $|g, 0\rangle$, is also an eigenstate of $\hat{H}_{\text{int}}$.

Upon introducing the operator:

$$\hat{N} = \hat{a}^\dagger \hat{a} + \frac{1}{2}\hat{\mathbb{1}} + \frac{1}{2}\hat{\sigma}_z, \quad (9.87)$$

the total Hamiltonian may be written as follows (after the RWA but not in the interaction picture):

$$\hat{H} = \hbar\omega \hat{N} + \frac{\hbar\delta}{2}\hat{\sigma}_z - i\frac{\hbar\Omega_0}{2}\left(\hat{\sigma}_+ \hat{a} - \hat{\sigma}_- \hat{a}^\dagger\right). \quad (9.88)$$

If we focus on the resonant case $\delta = 0$, besides the ground state, we find the following eigenstates of the total Hamiltonian for $n \geq 0$:

$$\hat{H}|\Psi_n^\pm\rangle = \hbar\underbrace{\left[(n+1)\omega \pm \frac{1}{2}\Omega_n\right]}_{E_n^\pm}|\Psi_n^\pm\rangle, \quad (9.89)$$

where:

$$|\Psi_n^\pm\rangle = \frac{1}{\sqrt{2}}\left(|e, n\rangle \pm i|g, n+1\rangle\right), \quad (9.90)$$

and $\Omega_n = \Omega_0\sqrt{n+1}$ is the Rabi frequency for n photons. The states $|\Psi_n^\pm\rangle$ are called *dressed states* and $\Delta E_n = E_n^+ - E_n^- = \hbar\Omega_0\sqrt{n+1}$. Of course we can also write:

$$|e, n\rangle = \frac{1}{\sqrt{2}}\left(|\Psi_n^+\rangle + |\Psi_n^-\rangle\right) \quad \text{and} \quad |g, n+1\rangle = \frac{1}{i\sqrt{2}}\left(|\Psi_n^+\rangle - |\Psi_n^-\rangle\right). \quad (9.91)$$

It is worth noting that the Jaynes–Cummings Hamiltonian of Eq. (9.84) can be also written as:

$$\hat{H}_{\text{int}} = \frac{\hbar\Omega_0}{2}\left(\hat{\sigma}_+ \hat{a} + \hat{\sigma}_- \hat{a}^\dagger\right), \quad (9.92)$$

where we perform the following unitary transformation of mode $\hat{a} \to i\hat{a}$, which, of course, preserves the commutation relations, since $[(i\hat{a}), (i\hat{a})^\dagger] = [\hat{a}, \hat{a}^\dagger] = \hat{\mathbb{1}}$. This transformation corresponds to apply a phase shift operator $\hat{U}_{\pi/2} = \exp(i\frac{\pi}{2}\hat{a}^\dagger \hat{a})$ and we have $\hat{U}_{\pi/2}^\dagger \hat{a} \hat{U}_{\pi/2} = i\hat{a}$.

9.5.1 Vacuum Rabi Oscillations: Quantum Circuit

If the atom is initially in the excited state $|e\rangle$ and the field is in the vacuum state $|0\rangle$, we have the *vacuum Rabi oscillations*. In particular we find:

- π-pulse ($\Omega_0 t = \pi$):

$$|e, 0\rangle \to |g, 1\rangle, \quad \text{and} \quad |g, 1\rangle \to -|e, 0\rangle; \tag{9.93}$$

- $\frac{\pi}{2}$-pulse ($\Omega_0 t = \pi/2$):

$$|e, 0\rangle \to \frac{1}{\sqrt{2}}(|e, 0\rangle + |g, 1\rangle), \quad \text{and} \quad |g, 1\rangle \to \frac{1}{\sqrt{2}}(|g, 1\rangle - |e, 0\rangle), \tag{9.94}$$

that are maximally entangled states of the atom and the cavity field.

The Fig. 9.3 shows how we can describe the vacuum Rabi oscillations by means of CNOT gates and controlled unitary operation:

$$\hat{R}(t) = \exp\left(-i\frac{\Omega_0 t}{2}\hat{\sigma}_y\right) = \cos\left(\frac{\Omega_0 t}{2}\right)\hat{\mathbb{1}} - i\sin\left(\frac{\Omega_0 t}{2}\right)\hat{\sigma}_y, \tag{9.95}$$

where we should use the following association between the physical states and the computational basis:

$$|g, 0\rangle \leftrightarrow |00\rangle, \quad |g, 1\rangle \leftrightarrow |01\rangle, \quad |e, 0\rangle \leftrightarrow |10\rangle, \quad \text{and} \quad |e, 1\rangle \leftrightarrow |11\rangle. \tag{9.96}$$

The reader can check that the quantum circuit of Fig. 9.3 acts on the computational basis as follows:

$$|00\rangle \to |00\rangle, \quad |11\rangle \to |11\rangle, \tag{9.97a}$$

$$|01\rangle \to \cos\left(\frac{\Omega_0 t}{2}\right)|01\rangle - \sin\left(\frac{\Omega_0 t}{2}\right)|10\rangle, \tag{9.97b}$$

$$|10\rangle \to \cos\left(\frac{\Omega_0 t}{2}\right)|10\rangle + \sin\left(\frac{\Omega_0 t}{2}\right)|01\rangle, \tag{9.97c}$$

Fig. 9.3 Quantum circuit implementing vacuum Rabi oscillations

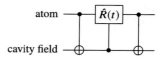

that is the same evolution obtained with the Jaynes–Cummings Hamiltonian (9.84), except for what concerns the state $|11\rangle = |e, 1\rangle$, since, in this case, we have:

$$\exp\left(-i\frac{\hat{H}_{\text{int}}}{\hbar}t\right)|e, 1\rangle = \cos\left(\frac{\Omega_1 t}{2}\right)|e, 1\rangle + \sin\left(\frac{\Omega_1 t}{2}\right)|g, 2\rangle. \tag{9.98}$$

As we have seen in the previous section, $\hat{H}_{\text{int}}$ couples the states $|e, 1\rangle$ and $|g, 2\rangle$, but $|g, 2\rangle$ does not belong to the computational space spanned by the two qubits...

In order solve this problem, we should modify the evolution as follows:

$$\exp\left(-i\frac{\hat{H}_{\text{int}}}{\hbar}t\right) \to \exp\left(-i\frac{\hat{H}_{\text{int}}}{\hbar}t\right)\left(\hat{P}_q - |e, 1\rangle\langle e, 1|\right) + |e, 1\rangle\langle e, 1|, \tag{9.99}$$

where we introduced the projector operator

$$\hat{P}_q = \sum_{A=g,e}\sum_{F=0,1} |A, F\rangle\langle A, F|, \tag{9.100}$$

which projects the state onto the 4-dimensional space spanned by the 2-qubit computational basis.

We close this section showing how we can map an atomic superposition state $|\psi_A\rangle = c_e|e\rangle + c_g|g\rangle$ onto the cavity field state. To this aim it is enough to prepare the field in the vacuum state and then apply a π-pulse, namely (note that, here, 0 and 1 represent the number of photons):

$$(c_e|e\rangle + c_g|g\rangle)|0\rangle \xrightarrow{\pi\text{-pulse}} |g\rangle(c_e|1\rangle + c_g|0\rangle), \tag{9.101}$$

i.e., the atom is left in the ground state while the cavity is a superposition state with the same complex amplitudes of the input atomic state. On the other hand, when we try to map the state $|\psi_A\rangle = c_1|1\rangle + c_0|0\rangle$ of the field onto an atomic state, we obtain:

$$|g\rangle(c_1|1\rangle + c_0|0\rangle) \xrightarrow{\pi\text{-pulse}} (-c_1|e\rangle + c_0|g\rangle)|0\rangle, \tag{9.102}$$

i.e., we have a phase appearing in front of $|e\rangle$. It is worth noting that the field considered throughout this chapter is *inside* a cavity and, thus, is not directly accessible: one should measure the atom after the interaction in order to have some information about the cavity state!

Problems

9.1. Starting form Eq. (9.9), explain why it is possible to reproduce the action of any single-qubit gate by using a single spin and a suitably chosen classical magnetic field.

9.2. Prove that the operators C_{12} and Z_{12} as defined in Eqs. (9.19) and (9.20b), respectively, act on $|x\rangle|y\rangle$ as a CNOT and a controlled-Z gates, where $\hat{\sigma}_z|x\rangle = (-1)^x|x\rangle$ and $\hat{\sigma}_x|x\rangle = |\bar{x}\rangle$.

9.3. Draw the quantum circuit to implement the CNOT gate involving spin-1/2 particles by using single-qubit gates and the two-qubit gate based on the exchange interaction. Explain how the involved magnetic fields should be directed, write their magnitude and the interaction time for each gate. Is it important to control the overall phases appearing on the quibit after the gates? Why?

9.4. Represent the evolution of the two-level atom interacting with a classical oscillating electric field by using the Bloch sphere formalism, in the case of $\frac{\pi}{2}$-pulse, π-pulse and 2π-pulse. Assume that the initial state is $|e\rangle$, that is the north pole of the unit sphere.

9.5. Assume that a two-level atom interacting with a quantized field system is initially prepared in the state $|e, n\rangle$, with $n \geq 0$. Calculate the probability to find the atom in the excited state after an interaction time t assuming $\delta = 0$.

9.6. Find the effect of a 2π-pulse ($\Omega_0 t = 2\pi$) on $|e, 0\rangle$ and $|g, 1\rangle$.

Further Readings

J. Stolze, D. Suter, *Quantum Computing: A Short Course from Theory to Experiment* (Wiley-VCH, Weinheim, 2004). Chapter 10

S. Haroche, J.-M. Raimond, *Exploring the Quantum: Atoms, Cavities, and Photons*. Oxford Graduate Texts (Oxford University Press, Oxford, 2006). Chapter 3, Chapter 5

M.A. Nielsen, I.L. Chuang, *Quantum Computation and Quantum Information* (Cambridge University Press, Cambridge, 2010). Chapter 7.5, Chapter 7.7

G. Burkard, D. Loss, D.P. DiVincenzo, J.A. Smolin, Physical optimization of quantum error correction circuits. Phys. Rev. B **60**, 11404–11416 (1999)

S. Olivares, Introduction to generation, manipulation and characterization of optical quantum states. Phys. Lett. A **418**, 127720 (2021)

E. Knill, R. Laflamme, G.J. Milburn, A scheme for efficient quantum computation with linear optics. Nature **409**, 46–52 (2001)

S. Aaronson, A. Arkhipov, The computational complexity of linear optics. Theory Comput. **9**, 143–252 (2013)

D. Gottesman, A. Kitaev, J. Preskill, Encoding a qubit in an oscillator. Phys. Rev. A **64**, 012310 (2001)

Quantum Computation with Trapped Ions

10

Abstract

In Chap. 9 we have seen how to manipulate a two-level atom by using oscillating electric field treated as classical or quantized entities. In that case the atoms usually move through a cavity which contains the field. A complementary approach consists in fixing the position of the atoms in the space and address suitably tuned laser beams in order to control their electronic levels and to perform quantum operations. In this last case, the atoms are ionized and *trapped* by using both static and time-varying electric fields: now one can exploit the electronic levels of the ions to encode the qubits' state, but also their collective quantized motion, that allows to implement two-qubit gates. In this chapter we review the basic working principle of a linear Paul trap, which is used to confine a chain of ions, we derive the quantum Hamiltonian describing their quantized motion and we investigate their manipulation through suitable classical laser pulses. We eventually show how to perform universal quantum computation with trapped ions.

10.1 The Linear Paul Trap (in Brief)

The typical linear Paul trap used to implement quantum computation consists of four rod electrodes which confine the ions in the (vertical) x-y plane, and two end-cap electrodes for the confinement along the (horizontal) z axis as depicted in Fig. 10.1. If we apply to one pair of the diagonally opposite electrodes a radio frequency (RF) voltage $V_1(t) = V \cos(\omega_{RF})$ and to the other couple of rod electrodes the voltage $V_2(t) = -V \cos(\omega_{RF})$, the time-varying potential along z axis (and near at the trap center) can be written as:

$$\Phi(x, y; t) = \phi_s(x, y) \cos(\omega_{RF} t). \tag{10.1}$$

© The Author(s), under exclusive license to Springer Nature Switzerland AG 2025
S. Olivares, *A Student's Guide to Quantum Computing*, Lecture Notes in Physics 1038, https://doi.org/10.1007/978-3-031-83361-8_10

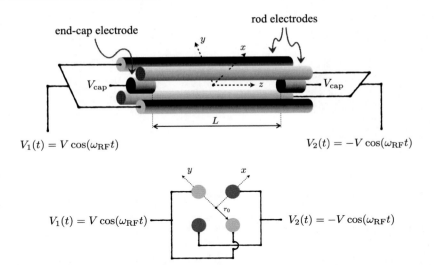

Fig. 10.1 Scheme of a linear Paul trap with its main elements. On the bottom we show a side view of the trap (the end-cap electrodes are not depicted). See the text for details

with

$$\phi_s(x, y) = V \frac{x^2 - y^2}{2r_0^2}, \tag{10.2}$$

where r_0 is the radial distance between the trap axis and the surface of one of the electrodes (see Fig. 10.1). This potential can be used to achieve radial confinement of charged particles.

If we consider a particle with mass m and charge Q, the classical equations of motion given the potential $\Phi(x, y; t)$ read:

$$\frac{d^2 x}{d\zeta^2} = 2q \cos(2\zeta) x, \tag{10.3a}$$

$$\frac{d^2 y}{d\zeta^2} = -2q \cos(2\zeta) y, \tag{10.3b}$$

$$\frac{d^2 z}{d\zeta^2} = 0, \tag{10.3c}$$

where we introduced the following dimensionless quantities:

$$q = \frac{2QV}{mr_0^2 \omega_{\mathrm{RF}}^2} \quad \text{and} \quad \zeta = \frac{\omega_{\mathrm{RF}} t}{2}. \tag{10.4}$$

10.1 The Linear Paul Trap (in Brief)

It is possible to show that the previous set of equations have *stable* solutions only if $0 < q < 0.908$: in this case the ion is confined radially, namely, in the x-y plane and can move freely along the z direction.

If we consider the x direction (an analogue result can be obtained for the y direction), the approximate solution can be written as:

$$x(t) \approx \underbrace{\mathcal{A}_x \cos(\omega_r t)}_{\text{secular motion}} \underbrace{\left[1 + \frac{q}{2} \cos(\omega_{RF} t)\right]}_{\text{micromotion}}, \qquad (10.5)$$

where the parameter $\mathcal{A}_x$ depends on the boundary condition, while:

$$\omega_{r_0} \equiv \frac{q\,\omega_{RF}}{2\sqrt{2}}. \qquad (10.6)$$

The micromotion can be eliminated by adding further electrodes operating with compensation voltages, therefore one can consider only the secular motion.

In order to confine the charged particles also in the z direction it is necessary to add the so-called end-cap electrodes, to which the same voltages V_{cap} is applied. In Fig. 10.1 we represented these electrodes as two rods placed on the trap axis. However, there are other possible geometries, such as ring-shaped electrodes around the RF-rods.

In the presence of the (DC) voltage V_{cap}, Eqs. (10.3) become:

$$\frac{d^2 x}{d\zeta^2} = 2q \cos(2\zeta)\, x - b, \qquad (10.7a)$$

$$\frac{d^2 y}{d\zeta^2} = -2q \cos(2\zeta)\, y - b, \qquad (10.7b)$$

$$\frac{d^2 z}{d\zeta^2} = -2bz, \qquad (10.7c)$$

where we introduced the new dimensionless parameter:

$$b = \alpha \frac{Q V_{\text{cap}}}{m L^2 \omega_{RF}^2}, \qquad (10.8)$$

α being a parameter depending on the geometry of the trap and L is the distance between the end-cap electrodes (see Fig. 10.1). From Eq. (10.7c) it is clear that now the particle exhibits a harmonic motion along z axis with frequency:

$$\omega_z = \sqrt{\frac{b}{2}}\, \omega_{RF}, \qquad (10.9)$$

while in the regime $b, q \ll 1$ the motion along the x direction (and, analogously, along the y axis) is still given by Eq. (10.5) but now the *pure* radial frequency ω_{r_0} should be replaced by:

$$\omega_r \approx \frac{\omega_{\text{RF}}}{2}\sqrt{\frac{q^2}{2}-b} \approx \sqrt{\omega_{r_0}^2 - \frac{\omega_z^2}{2}}. \tag{10.10}$$

Therefore, we find a defocusing effect of the radial motion, due to the confinement along the trap axis. Nevertheless, in the cases of interest one chooses the regime $\omega_z \ll \omega_{r_0}$, thus the defocusing can be safely neglected.

Summarizing, all the above considerations allow us to describe the trapped ion as a charged particle confined in a 3-dimensional harmonic potential, namely:

$$\Xi_1(x, y, z) = \frac{m}{2}\left[\omega_r^2(x^2+y^2) + \omega_z^2 z^2\right], \tag{10.11}$$

where we assumed $\omega_x = \omega_y = \omega_r$, that is the two radial frequencies are degenerate.

10.2 The Ion Chain

In order to perform quantum information tasks, one should manipulate more than one ion at the time. Therefore, we should extend our analysis to N charged particles. In the following we assume that all the ions have the same mass m and charge Q and, taking into account the mutual Coloumb interactions, we obtain the following potential:

$$\Xi_N(x, y, z) = \frac{m}{2}\sum_{n=1}^{N}\left[\omega_r^2(x_n^2+y_n^2) + \omega_z^2 z_n^2\right] + \frac{Q^2}{8\pi\epsilon_0}\sum_{\substack{n,m=1 \\ n\neq m}}^{N}\frac{1}{|\mathbf{r}_n-\mathbf{r}_m|}, \tag{10.12}$$

$\mathbf{r}_n = (x_n, y_n, z_n)$ being the position vector of the n-th ion.

If the radial confinement is stronger enough than the axial one and the number N is not too large, we obtain a linear ion chain configuration, in which the equilibrium positions of the ions are along the trap axis. This configuration is called ion crystal.

In general, the distance between adjacent ions increases from the center to the outside of the string, and can be evaluated by numerical calculations. However, by increasing the number of ions we find a transition from the linear chain to the so-called zig-zag configuration (or other more complicated ones). The value of N, above which the transition occurs, has been investigated both numerically and experimentally and one finds the following condition on the ration of the involved frequencies:

$$\mathcal{R} \equiv \left(\frac{\omega_z}{\omega_r}\right)^2 \lesssim 2.53\, N^{-1.73} \equiv \mathcal{R}_{\text{crit}}. \tag{10.13}$$

If $\mathcal{R} < \mathcal{R}_{\text{crit}}$ the zig-zag motion is suppressed and we can focus on the axial motion of the particles. Under this condition, we study the dynamics of our system given the potential:

$$\xi_N(z) = \frac{m}{2} \sum_{n=1}^{N} \omega_z^2 z_n^2 + \frac{Q^2}{8\pi\epsilon_0} \sum_{\substack{n,m=1 \\ n\neq m}}^{N} \frac{1}{|z_n - z_m|}. \tag{10.14}$$

The thorough investigation of the dynamics of the N-ion chain is beyond the scope of this chapter. Here we recall that we can identify two main axial modes. The first mode corresponds to the center-of-mass (COM) axial mode, where all the ions move along the z direction with the same amplitude and frequency ω_z. The second mode is the breathing mode: in this case the amplitude of oscillation of each ion increases as the distance from the center becomes larger.

In the following, we will assume that our system is excited in the COM axial mode and we represent the position of the n-th ion as follows:

$$z_n(t) = \bar{z}_n + \Delta_n(t), \tag{10.15}$$

where $\bar{z}_n$ is its average equilibrium position and $\Delta_n(t)$ its time-dependent displacement. We note that it is possible to impose the normal oscillation modes by acting with an AC voltage on the end-cap electrodes. In the next section we will describe the motion of the ions as a *quantum* harmonic oscillator.

10.3 Quantum Motion of the Ion Chain

If we consider just two electronic levels of each ion with transition frequency ω_A and assume the COM axial mode at frequency ω_z, the free quantum Hamiltonian of the system can be written as:

$$\hat{H}_0 = \sum_{n=1}^{N} \frac{\hbar \omega_A}{2} \hat{\sigma}_z^{(n)} + \hbar \omega_z \left(\hat{a}^\dagger \hat{a} + \frac{1}{2} \right), \tag{10.16}$$

where we introduced the annihilation, $\hat{a}$, and creation, $\hat{a}^\dagger$, operators of the harmonic oscillator, $[\hat{a}, \hat{a}^\dagger] = \hat{\mathbb{1}}$. The position z_n is then substituted by the operator:

$$\hat{z}_n = \bar{z}_n + \frac{z_0}{\sqrt{N}} \left(\hat{a} + \hat{a}^\dagger \right), \tag{10.17}$$

with

$$z_0 = \sqrt{\frac{\hbar}{2m\omega_z}}. \tag{10.18}$$

In order to manipulate the internal levels of the n-th ion at position $\hat{r}_n = (0, 0, \hat{z}_n)$, one should address on it a laser beam, whose electric filed writes (without loss of generality we assume a real polarization vector ε_f)

$$E_n(t) = E_0\, \varepsilon_f \left[e^{-i(\omega t - \mathbf{k}\cdot \mathbf{r}_n - \varphi)} + e^{i(\omega t - \mathbf{k}\cdot \mathbf{r}_n - \varphi)} \right], \qquad (10.19)$$

ω, $\mathbf{k}$ and φ being the laser frequency, the wave vector and the phase of the electric field, respectively. The interaction Hamiltonian can be written in terms of both the dipole and quadrupole operators and reads:

$$\hat{H}_{\text{int}} = -\hat{\mathbf{D}}_n \cdot E_0\, \varepsilon_f \left\{ e^{-i[\omega t - \eta(\hat{a}+\hat{a}^\dagger) - \varphi_n]} + e^{i[\omega t - \eta(\hat{a}+\hat{a}^\dagger) - \varphi_n]} \right\} \qquad (10.20)$$

where

$$\hat{\mathbf{D}}_n = d_n \varepsilon_a \left(\hat{\sigma}_+^{(n)} + \hat{\sigma}_-^{(n)} \right) \qquad (10.21)$$

is the dipole moment operator of the n-th ion (we assume $\varepsilon_a \in \mathbb{R}^3$), $\hat{\sigma}_+^{(n)}$ and $\hat{\sigma}_-^{(n)}$ are its raising and lowering operators, respectively (see Sect. 9.2), and:

$$\varphi_n = \varphi - |\mathbf{k}|\, \bar{z}_n \cos\theta, \qquad (10.22)$$

θ being the angle between the wave vector $\mathbf{k}$ and the z axis (see Fig. 10.2). In Eq. (10.20) we introduced the Lamb–Dicke parameter:

$$\eta = \frac{1}{\sqrt{N}}\, |\mathbf{k}|\, z_0 \cos\theta. \qquad (10.23)$$

Now we pass to the interaction picture with respect to the free Hamiltonian (10.16) and perform the RWA, obtaining the following Hamiltonian (see Appendix A):

$$\hat{H}'_{\text{int}} = -\frac{\hbar \Omega_0}{2} \left\{ \hat{\sigma}_+^{(n)} e^{-i\delta t} \exp\left[i\eta \left(\hat{a}\, e^{-i\omega_z t} + \hat{a}^\dagger e^{i\omega_z t} \right) + i\varphi_n \right] \right.$$
$$\left. + \hat{\sigma}_-^{(n)} e^{i\delta t} \exp\left[-i\eta \left(\hat{a}\, e^{-i\omega_z t} + \hat{a}^\dagger e^{i\omega_z t} \right) - i\varphi_n \right] \right\}, \qquad (10.24)$$

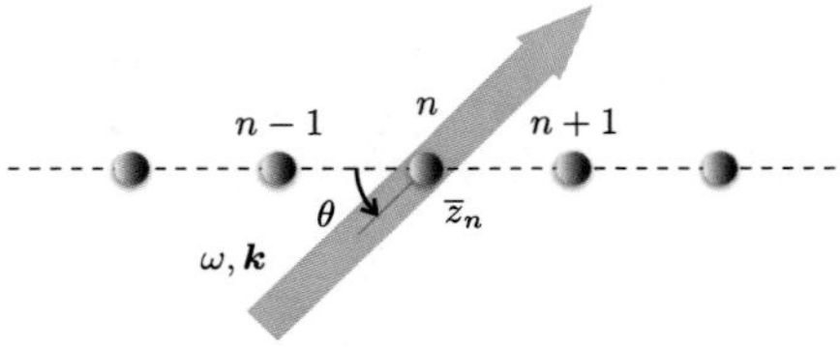

Fig. 10.2 Sketch of the interaction of a laser beam with frequency ω and wave vector $\mathbf{k}$ with the n-th ion

10.3 Quantum Motion of the Ion Chain

where $\delta = \omega - \omega_A$ is the laser-ion detuning and:

$$\Omega_0 = \frac{2d_n E_0}{\hbar} \, \boldsymbol{\varepsilon}_a \cdot \boldsymbol{\varepsilon}_f \tag{10.25}$$

is the Rabi frequency. If we consider the so-called Lamb–Dicke regime, namely:

$$\eta^2 \langle (\hat{a} + \hat{a}^\dagger)^2 \rangle = \eta^2 (2\bar{n} + 1) \ll 1, \tag{10.26}$$

where $\bar{n}$ is the average number of *phonons* and the expectation is calculated considering the COM mode state, we can expand $\hat{H}'_{\text{int}}$ up to the first order in η, obtaining:

$$\hat{H}'_{\text{int}} \approx -\frac{\hbar\Omega_0}{2} \left\{ \hat{\sigma}_+^{(n)} e^{-i\delta t + i\varphi_n} \left[1 + i\eta \left(\hat{a} e^{-i\omega_z t} + \hat{a}^\dagger e^{i\omega_z t} \right) \right] \right.$$
$$\left. + \hat{\sigma}_-^{(n)} e^{i\delta t - i\varphi_n} \left[1 - i\eta \left(\hat{a} e^{-i\omega_z t} + \hat{a}^\dagger e^{i\omega_z t} \right) \right] \right\}. \tag{10.27}$$

In order to perform universal quantum computation with the trapped ions we choose three particular values of the detuning δ. If we set $\delta = 0, \pm\omega_z$ and we neglect the oscillating terms $e^{\pm i\omega_z t}$ and $e^{\pm 2i\omega_z t}$, we obtain the following three Hamiltonians:

$$\hat{H}_C = -\frac{\hbar\Omega_0}{2} \left(\hat{\sigma}_+^{(n)} e^{i\varphi_n} + \hat{\sigma}_-^{(n)} e^{-i\varphi_n} \right), \tag{10.28a}$$
$$(\delta = 0, \text{ carrier})$$

$$\hat{H}_B = -i\eta \frac{\hbar\Omega_0}{2} \left(\hat{\sigma}_+^{(n)} \hat{a}^\dagger e^{i\varphi_n} - \hat{\sigma}_-^{(n)} \hat{a} e^{-i\varphi_n} \right), \tag{10.28b}$$
$$(\delta = +\omega_z, \text{ first blue sideband})$$

$$\hat{H}_R = -i\eta \frac{\hbar\Omega_0}{2} \left(\hat{\sigma}_+^{(n)} \hat{a} e^{i\varphi_n} - \hat{\sigma}_-^{(n)} \hat{a}^\dagger e^{-i\varphi_n} \right). \tag{10.28c}$$
$$(\delta = -\omega_z, \text{ first red sideband})$$

In Fig. 10.3 we sketch the allowed transition in the presence of the three Hamiltonians (10.28). We see that by suitably tuning the laser frequency ω one can obtain a transition between the electronic levels of the n-th ion preserving the phonon number state $|m\rangle$, namely:

$$|g_n\rangle|m\rangle \leftrightarrow |e_n\rangle|m\rangle, \tag{10.29}$$
$$(\text{carrier transition})$$

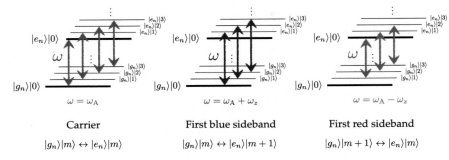

Fig. 10.3 Scheme of the allowed transition between the electronic ($|g_n\rangle$ and $|e_n\rangle$) and vibrational levels ($|m\rangle$) of the n-th ion in the presence of the carrier, first blue sideband and first red sideband transitions, respectively

or change the internal level and adding one vibrational quantum:

$$|g_n\rangle|m\rangle \leftrightarrow |e_n\rangle|m+1\rangle, \quad (10.30)$$
$$\text{(first blue sideband transition)}$$

or removing one vibrational quantum:

$$|g_n\rangle|m+1\rangle \leftrightarrow |e_n\rangle|m\rangle. \quad (10.31)$$
$$\text{(first red sideband transition)}$$

In the next sections we will see that exploiting the three considered Hamiltonians it is possible to implement universal quantum computation with trapped ions. To this aim, the logical qubits are encoded into the electronic levels of the ions and the COM mode is used as a bus to perform multi-qubit conditional operations.

10.4 Single-Qubit Gates with Trapped Ions

If we identify the computational basis $|0_n\rangle$ and $|1_n\rangle$ with the level $|g_n\rangle$ and $|e_n\rangle$, of the n-th ion, respectively, we can implement single qubit gates exploiting the carrier transition and following the analysis given in Sect. 9.2.1. The carrier Hamiltonian (10.28a) leads to the evolution operator (acting on the n-th ion):

$$\hat{C}_n(\theta,\varphi) = \exp\left[-i\frac{\theta}{2}\left(\hat{\sigma}_+^{(n)} e^{i\varphi} + \hat{\sigma}_-^{(n)} e^{-i\varphi}\right)\right], \quad (10.32)$$

$$= \cos\left(\frac{\theta}{2}\right)\hat{\mathbb{I}} - i\sin\left(\frac{\theta}{2}\right)\left(\cos\varphi\,\hat{\sigma}_x^{(n)} - \sin\varphi\,\hat{\sigma}_y^{(n)}\right) \quad (10.33)$$

where $\theta = \Omega_0 t$ and we used $\hat{\sigma}_n^{(\pm)} = \frac{1}{2}\left(\hat{\sigma}_x^{(n)} \pm i\hat{\sigma}_y^{(n)}\right)$. In particular, we obtain the following relevant cases that will be used in the next section to implement the CNOT gate:

$$\hat{C}_n\left(\frac{\pi}{2}, 0\right)|g_n\rangle = \frac{|g_n\rangle - i|e_n\rangle}{\sqrt{2}}, \quad \hat{C}_n\left(\frac{\pi}{2}, 0\right)|e_n\rangle = \frac{|e_n\rangle - i|g_n\rangle}{\sqrt{2}}, \quad (10.34\text{a})$$

$$\hat{C}_n\left(\frac{\pi}{2}, \pi\right)|g_n\rangle = \frac{|g_n\rangle + i|e_n\rangle}{\sqrt{2}}, \quad \hat{C}_n\left(\frac{\pi}{2}, \pi\right)|e_n\rangle = \frac{|e_n\rangle + i|g_n\rangle}{\sqrt{2}}. \quad (10.34\text{b})$$

In order to achieve the universal quantum computation with trapped ions, we now need to build the CNOT operation, that will be the subject of the next section.

10.5 CNOT Gate with Trapped Ions

In this section, inspired by the Cirac–Zoller CNOT gate, we will consider two particular ions of an ion chain, say ion 1 and ion 2, and we will exploit the common COM axial mode to change the electronic state of the ion 2 only if the ion 1 is in the excited state $|e_1\rangle$. Therefore, if we use as computational basis:

$$|g_n\rangle \to |0_n\rangle, \quad \text{and} \quad |e_n\rangle \to |1_n\rangle, \quad (10.35)$$

the final result is the action of a CNOT gate (up to global phases), that is:

$$|g_1\rangle|g_2\rangle = |0_1\rangle|0_2\rangle \to |g_1\rangle|g_2\rangle = |0_1\rangle|0_2\rangle, \quad (10.36\text{a})$$

$$|g_1\rangle|e_2\rangle = |0_1\rangle|1_2\rangle \to |g_1\rangle|e_2\rangle = |0_1\rangle|1_2\rangle, \quad (10.36\text{b})$$

$$|e_1\rangle|g_2\rangle = |1_1\rangle|0_2\rangle \to |e_1\rangle|e_2\rangle = |1_1\rangle|1_2\rangle, \quad (10.36\text{c})$$

$$|e_1\rangle|e_2\rangle = |1_1\rangle|1_2\rangle \to |e_1\rangle|g_2\rangle = |1_1\rangle|0_2\rangle. \quad (10.36\text{d})$$

In order to implement the conditional operations needed to obtain the action of the CNOT gate, we will use the collective motion imposed by the COM mode by applying suitable carrier and first blue sideband pulses to the ions. For the sake of simplicity, we introduce the following evolution operator associated with the Hamiltonian (10.28b):

$$\hat{B}_n(\theta, \varphi) = \exp\left[-i\frac{\theta}{2}\left(\hat{\sigma}_+^{(n)}\hat{a}^\dagger e^{i\varphi} + \hat{\sigma}_-^{(n)}\hat{a}\, e^{-i\varphi}\right)\right], \quad (10.37)$$

where $\theta = \eta\Omega_0 t$ and we applied the transformation $\hat{a} \to i\hat{a}$. As in the case of the Jaynes–Cummings model described in Sect. 9.5, we have:

$$\left(\hat{\sigma}_+^{(n)}\hat{a}^\dagger e^{i\varphi} + \hat{\sigma}_-^{(n)}\hat{a}\, e^{-i\varphi}\right) \left|\Psi_{n,m}^{(\pm)}\right\rangle = \pm\sqrt{m+1}\left|\Psi_{n,m}^{(\pm)}\right\rangle, \tag{10.38}$$

where:

$$\left|\Psi_{n,m}^{(\pm)}\right\rangle = \frac{|g_n\rangle|m\rangle \pm e^{i\varphi}|e_n\rangle|m+1\rangle}{\sqrt{2}}, \tag{10.39}$$

$|m\rangle$ being the phonon Fock state, $\hat{a}^\dagger\hat{a}|m\rangle = m|m\rangle$.

It is straightforward to show that (see problem 10.1):

$$\hat{B}_n(\theta,\varphi)|g_n\rangle|m\rangle = \cos\left(\frac{\theta}{2}\sqrt{m+1}\right)|g_n\rangle|m\rangle$$

$$- ie^{i\varphi}\sin\left(\frac{\theta}{2}\sqrt{m+1}\right)|e_n\rangle|m+1\rangle, \tag{10.40a}$$

$$\hat{B}_n(\theta,\varphi)|e_n\rangle|m+1\rangle = \cos\left(\frac{\theta}{2}\sqrt{m+1}\right)|e_n\rangle|m+1\rangle$$

$$- ie^{-i\varphi}\sin\left(\frac{\theta}{2}\sqrt{m+1}\right)|g_n\rangle|m\rangle. \tag{10.40b}$$

In Fig. 10.4 we show the quantum circuit to implement a CNOT gate with trapped ions, that is a suitable combination of carrier and first blue sideband pulses applied to the two involved ions. As one can see there are difference pulses which are required to control the phases raising from the first blue sideband gates. In the considered case, the CNOT gate uses the ion 1 as control qubit and changes the internal (electronic) state of ion 2, the target state, only in the presence of the state $|e_1\rangle$.

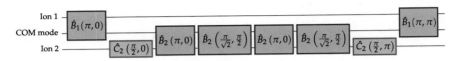

Fig. 10.4 Quantum circuit used to implement a CNOT gate that changes the internal state of the ion 2 only if the ion 1 is in the excited state. The gates $\hat{C}_n(\theta,\varphi)$ and $\hat{B}_n(\theta,\varphi)$ refer to the carrier and to the first blue sideband pulses, respectively, as mentioned in the text. The COM mode is used as a bus

10.5 CNOT Gate with Trapped Ions

To understand the basic idea underlying the circuit of Fig. 10.4, we note that the first gate, $\hat{B}_1(\pi, 0)$, maps the internal state of the first ion into the COM axial mode. In fact, if the starting state of the COM mode is $|0\rangle$, we have:

$$\hat{B}_1(\pi, 0)|g_1\rangle|0\rangle = -i|e_1\rangle|1\rangle \quad \text{and} \quad \hat{B}_1(\pi, 0)|e_1\rangle|0\rangle = |e_1\rangle|0\rangle, \quad (10.41)$$

and we can see how the state of the COM mode is changed to $|1\rangle$ only if the ion 1 is in its ground state $|g_1\rangle$. All the other $\hat{C}_2(\theta, \phi)$ and $\hat{B}_2(\theta, \phi)$ gates are used to manipulate the state of the ion 2.

As an example, we consider the whole evolution of the state initial state

$$|e_1\rangle|e_2\rangle|0\rangle \equiv |1_1\rangle|1_2\rangle|0\rangle, \quad (10.42)$$

where we used both the physical basis (l.h.s.) and the computational basis (r.h.s.) for the ion states. Since, $\hat{B}_1(\pi, 0)|e_1\rangle|0\rangle = |e_1\rangle|0\rangle$, we can focus on the evolution of $|e_2\rangle|0\rangle$. Following the circuit in Fig. 10.4, we have:

$$|e_2\rangle|0\rangle \xrightarrow{\hat{C}_2(\frac{\pi}{2},0)} \frac{|e_2\rangle - i|g_2\rangle}{\sqrt{2}}|0\rangle \quad (10.43a)$$

$$\xrightarrow{\hat{B}_2(\pi,0)} |e_2\rangle\frac{|0\rangle - |1\rangle}{\sqrt{2}} \quad (10.43b)$$

$$\xrightarrow{\hat{B}_2\left(\frac{\pi}{\sqrt{2}},\frac{\pi}{2}\right)} \frac{1}{\sqrt{2}}\left[|e_2\rangle|0\rangle - \cos\left(\frac{\pi}{2\sqrt{2}}\right)|e_2\rangle|1\rangle + \sin\left(\frac{\pi}{2\sqrt{2}}\right)|g_2\rangle|0\rangle\right] \quad (10.43c)$$

$$\xrightarrow{\hat{B}_2(\pi,0)} \frac{1}{\sqrt{2}}\left[|e_2\rangle|0\rangle + i\cos\left(\frac{\pi}{2\sqrt{2}}\right)|g_2\rangle|0\rangle - i\sin\left(\frac{\pi}{2\sqrt{2}}\right)|e_2\rangle|1\rangle\right] \quad (10.43d)$$

$$\xrightarrow{\hat{B}_2\left(\frac{\pi}{\sqrt{2}},\frac{\pi}{2}\right)} \frac{|e_2\rangle + i|b_2\rangle}{\sqrt{2}}|0\rangle \quad (10.43e)$$

$$\xrightarrow{\hat{C}_2\left(\frac{\pi}{2},\pi\right)} i|g_2\rangle|0\rangle, \quad (10.43f)$$

and, thus, the output state after the whole circuit is (up to a global phase):

$$|e_1\rangle|g_2\rangle|0\rangle \equiv |1_1\rangle|0_2\rangle|0\rangle, \quad (10.44)$$

where we used $\hat{B}_1(\pi, \pi)|e_1\rangle|0\rangle = |e_1\rangle|0\rangle$. Therefore, we get:

$$|1_1\rangle|1_2\rangle \to |1_1\rangle|0_2\rangle, \quad (10.45)$$

as expected. Analogous results can be obtained for the other two-ion states (see problem 10.2).

In conclusion, we have shown the possibility to implement a CNOT gate. This result, together with the single ion operations described in Sect. 10.4, proves that it is possible to perform universal quantum computation with trapped ions.

10.6 Hyperfine and Optical Qubits

There are two possible ways to actually implement a qubit with trapped ions, which require different species of ions according to the presence or not of the nuclear angular momentum.

Ions such as $^9\text{Be}^+$, $^{43}\text{Ca}^+$ and $^{171}\text{Yb}^+$ exhibit non-zero nuclear angular momentum. Here the logical levels are the hyperfine structure of the ground state and the frequencies involved are of the order of GHz (microwaves).

In the case of ions with zero nuclear angular momentum, like $^{40}\text{Ca}^+$, $^{88}\text{Sr}^+$ and $^{174}\text{Yb}^+$, the logical levels are obtained within the fine structure and a metastable excited state. Now we have optical frequencies with quadrupole transition leading to a longer state's lifetime.

Problems

10.1. Prove Eqs. (10.40) by performing the explicit calculations using Eqs. (10.38) and (10.39).

10.2. Prove that the quantum circuit represented in Fig. 10.4 acts a CNOT gate for the ion states (up to a global phase) and, in particular, one has:

$$|g_1\rangle|g_2\rangle|0\rangle \rightarrow -|g_1\rangle|g_2\rangle|0\rangle,$$

$$|g_1\rangle|e_2\rangle|0\rangle \rightarrow -|g_1\rangle|e_2\rangle|0\rangle,$$

$$|e_1\rangle|g_2\rangle|0\rangle \rightarrow -i|e_1\rangle|e_2\rangle|0\rangle,$$

$$|e_1\rangle|e_2\rangle|0\rangle \rightarrow i|e_1\rangle|g_2\rangle|0\rangle.$$

Further Readings

- D. Leibfried, R. Blatt, C. Monroe, D. Wineland, Quantum dynamics of single trapped ions. Rev. Mod. Phys. **75**, 281–324 (2003)
- S. Gulde, Experimental realization of quantum gates and the Deutsch–Jozsa algorithm with trapped $^{40}Ca^+$ ions. Ph.D. Thesis, Leopold-Franzens-Universität, Innsbruck (2003)
- F. Schmidt-Kaler, H. Häffner, M. Riebe, S. Gulde, G. Lancaster, T. Deuschle, C. Becher, C. Roos, J. Eschner, R. Blatt, Realization of the Cirac–Zoller controlled-NOT quantum gate. Nature **422**, 408–411 (2003)

Superconducting Qubits: Charge and Transmon Qubits

Abstract

In this chapter we explain how to obtain a two-level system starting from superconducting circuits. In particular we consider the Josephson junction and the SQUID (Superconducting QUantum Interference Device) and we focus on the charge qubit and the transmon qubit. We also describe the coupling between a charge qubit and a 1-D transmission line resonator leading to a coupling Hamiltonian similar to that obtained in cavity QED experiments.

11.1 The LC Circuit as a Quantum Harmonic Oscillator

We consider a circuit involving an inductor, with inductance L, and and a capacitor, with capacity C. If we indicate with V the voltage at the ends of the capacitor and with I the current flowing in the circuit, the energies stored in the capacitor and in the inductor are:

$$E_C = \frac{1}{2}CV^2 = \frac{Q^2}{2C}, \quad \text{and} \quad E_L = \frac{1}{2}LI^2 = \frac{\Phi^2}{2L}, \tag{11.1}$$

respectively, where $Q = CV$ is the charge of the capacitor and $\Phi = LI$ is the magnetic flux in the inductor. The classical Hamiltonian of this system, $H_{\text{cl}} = E_C + E_L$, explicitly reads:

$$H_{\text{cl}} = \frac{Q^2}{2C} + \frac{\Phi^2}{2L}, \tag{11.2a}$$

$$= \frac{Q^2}{2C} + \frac{1}{2}C\omega_0^2 \Phi^2, \tag{11.2b}$$

© The Author(s), under exclusive license to Springer Nature Switzerland AG 2025
S. Olivares, *A Student's Guide to Quantum Computing*, Lecture Notes in Physics 1038, https://doi.org/10.1007/978-3-031-83361-8_11

that is the classical Hamiltonian of a harmonic oscillator with "mass" C, "momentum" Q, "position" Φ and frequency $\omega_0 = 1/\sqrt{LC}$.

11.1.1 Quantization of the LC Circuit

The quantization of H_{cl} is achieved by the substitution (see also Sect. 9.3):

$$Q \to \hat{Q} = i\sqrt{\frac{\hbar}{2Z_0}}\left(\hat{a}^\dagger - \hat{a}\right), \quad \text{and} \quad \Phi \to \hat{\Phi} = \sqrt{\frac{\hbar Z_0}{2}}\left(\hat{a}^\dagger + \hat{a}\right), \quad (11.3)$$

where we introduced the impedance $Z_0 = \sqrt{L/C}$ and the annihilation and creation operators $\hat{a}$ and $\hat{a}^\dagger$, respectively, $[\hat{a}, \hat{a}^\dagger] = \hat{\mathbb{1}}$. Note that $\hat{\Phi}$ and $\hat{Q}$ are conjugated quantum variables, namely:

$$\left[\hat{\Phi}, \hat{Q}\right] = i\hbar\hat{\mathbb{1}}. \quad (11.4)$$

Thereafter, the quantum Hamiltonian associated with the LC circuit writes:

$$\hat{H}_{LC} = \hbar\omega_0\left(\hat{a}^\dagger\hat{a} + \frac{1}{2}\right) \quad (11.5)$$

and:

$$\hat{H}_{LC}|n\rangle = E_n|n\rangle, \quad (11.6)$$

where $|n\rangle$, $n \in \mathbb{N}$, are the corresponding eigenstates with eigenvalues:

$$E_n = \hbar\omega_0\left(n + \frac{1}{2}\right). \quad (11.7)$$

Since the difference between two levels $\Delta E = E_{n+1} - E_n = \hbar\omega_0$ is independent of n, we cannot select *only* two particular levels in order to realize a qubit. To make the energies of the quantized levels different enough to obtain a two-level system, we should introduce some nonlinearity, which leads to a nonlinear oscillator.

11.2 The Josephson Junction and the SQUID

A Josephson junction consists of two superconductors connected via a tunnelling barrier. It can be described by its critical current I_c, and the *gauge invariant phase difference* φ across the junction. The actual value of the critical current depends on the superconducting material and the size of the junction. Though the analysis of this system requires the knowledge of the superconductivity theory, in the following

11.2 The Josephson Junction and the SQUID

we provide the reader with the few theoretical aspects useful to better understand the physics underlying the Josephson junction.

We associate with each superconductor $k = 1, 2$ the wave function:

$$\Psi_k = \sqrt{\rho_k}\, e^{i\phi_k}, \qquad (11.8)$$

where ρ_k is the density of Cooper pairs of the k-th and ϕ_k its phase. The dynamics of the system is then described by the coupled Schrödinger equations:

$$i\hbar \frac{\partial \Psi_1}{\partial t} = E_1 \Psi_1 + \kappa \Psi_2, \qquad (11.9a)$$

$$i\hbar \frac{\partial \Psi_2}{\partial t} = E_2 \Psi_2 + \kappa \Psi_1, \qquad (11.9b)$$

where E_1 and E_2 are the energies of the states and κ the coupling constant which measures the interaction of the two wave functions. By substituting the expression of Ψ_k into the Schrödinger equations we find the following equations:

$$\hbar \frac{\partial \rho_1}{\partial t} = 2\kappa \sqrt{\rho_1 \rho_2}\, \sin\varphi, \qquad (11.10a)$$

$$\hbar \frac{\partial \rho_2}{\partial t} = -2\kappa \sqrt{\rho_1 \rho_2}\, \sin\varphi, \qquad (11.10b)$$

$$\hbar \frac{\partial \phi}{\partial t} = E_2 - E_1. \qquad (11.10c)$$

The derivatives $\partial_t \rho_1 = -\partial_t \rho_2$ are proportional to the so-called Josephson current I_J, while the quantity $2\kappa \sqrt{\rho_1 \rho_2}$ to critical current I_c mentioned above. Moreover, if we apply a voltage V to the junction, we have $E_2 - E_1 = 2eV$, e being the charge of the electron, and the previous equations can be rewritten as the two following *Josephson equations*:

$$I_J(t) = I_c \sin\varphi(t), \qquad \text{(1st Josephson equation)} \qquad (11.11)$$

$$\frac{\partial \varphi(t)}{\partial t} = \frac{2\pi}{\Phi_0} V, \qquad \text{(2nd Josephson equation)} \qquad (11.12)$$

that allow to describe the time evolution of the Josephson current I_J and of φ as a function of the applied voltage V. In Eq. (11.12) we introduced the *superconducting flux quantum*:

$$\Phi_0 = \frac{h}{2e} = 2.07 \times 10^{-15}\ \text{Wb}, \qquad (11.13)$$

where $2e$ is the charge of a *Cooper pair*. The time derivative of Eq. (11.11) gives:

$$i_J = I_c \cos\varphi \, \frac{\partial \varphi}{\partial t}, \tag{11.14}$$

and, using Eq. (11.12) and since $\dot{I} = V/L$, we can introduce the following *nonlinear inductance*:

$$L_J = \frac{1}{\cos\varphi} \frac{\Phi_0}{2\pi I_c}. \tag{11.15}$$

The energy associated with L_J is obtained as follows:

$$E_J^{(L)} = \int_0^t d\tau \, I_J(\tau) \, V = E_J(1 - \cos\varphi), \tag{11.16}$$

where:

$$E_J = \frac{\Phi_0 I_c}{2\pi} \tag{11.17}$$

is the Josephson energy, which is a measure of the coupling across the junction. Since a Josephson junction has also a capacitance C_J, we can calculate the corresponding energy:

$$E_J^{(C)} = \frac{Q^2}{2C_J}, \tag{11.18}$$

where Q is the charge of the junction (seen as a capacitor).

The classical Hamiltonian of the Josephson junction can be written as (we neglect the constant term E_J):

$$H_J = \frac{Q^2}{2C_J} - E_J \cos\varphi. \tag{11.19}$$

Since $Q = (2e)N$, where $N \in (-\infty, +\infty)$ is the *excess* of Cooper pair in the junction, $N = N_1 - N_2$, where N_1 and N_2 represent the numbers of Cooper pairs present at each side of the junction, we can define the capacitive energy:

$$E_c = \frac{e^2}{2C_J}, \tag{11.20}$$

and Eq. (11.19) becomes:

$$H_J = 4E_c N^2 - E_J \cos\varphi. \tag{11.21}$$

11.2 The Josephson Junction and the SQUID

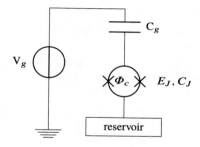

Fig. 11.1 A SQUID embedded in a circuit with a gate voltage V_g

Instead of a single Josephson junction we can consider two Josephson junctions connected in parallel on a superconducting loop: this system is called superconducting quantum interference device (SQUID). If the inductance of the loop can be neglected, then the corresponding Hamiltonian is the same as in Eq. (11.21), but now:

$$C_J \to 2C_J^{(s)}, \tag{11.22}$$

$$E_J \to E_J(\Phi_c) = 2E_J^{(s)} \cos\left(\pi \frac{\Phi_c}{\Phi_0}\right), \tag{11.23}$$

where $C_J^{(s)}$ and $E_J^{(s)}$ are the single Josephson junction capacitance and energy, respectively, and Φ_c is the (eventual) external flux: changing Φ_c one can modify E_J.

From now on we assume that our system is a SQUID embedded in a circuit and a gate voltage V_g is applied through a capacitance C_g, as shown in Fig. 11.1. The presence of V_g simply shifts N in Eq. (11.21) by:

$$N_g = \frac{C_g V_g}{2e}, \tag{11.24}$$

namely:

$$H = 4E_c(N - N_g)^2 - E_J \cos\varphi, \tag{11.25}$$

where, now:

$$E_c = \frac{e^2}{2(C_J + C_g)}. \tag{11.26}$$

If we associate $4E_c N^2$ with the kinetic energy and $-E_J \cos\varphi$ with the potential energy, then Eq. (11.25) represents the Hamiltonian of a nonlinear oscillator, where the conjugated variables are N and φ, corresponding to the momentum and to the position, respectively.

11.2.1 Quantization of the Josephson Junction and SQUID Hamiltonians

We can now obtain the quantum analogue of the Hamiltonian Eq. (11.25) associating with φ and N the corresponding quantum operators:

$$\varphi \to \hat{\varphi}, \quad N \to \hat{N}, \tag{11.27}$$

and the quantum Hamiltonian reads:

$$\hat{H} = 4E_c(\hat{N} - N_g)^2 - E_J \cos\hat{\varphi}. \tag{11.28}$$

It is worth noting that $\hat{N}$ is the operator associated with the excess of Cooper pairs N, where $N \in (-\infty, +\infty)$, and does not correspond to the *number operator* of the quantum harmonic oscillator, as the one considered for the electromagnetic field in Sect. 9.3. We can write the relation between $\hat{\varphi}$ and $\hat{N}$ as:

$$e^{i\hat{\varphi}} \hat{N} e^{-i\hat{\varphi}} = \hat{N} - \hat{\mathbb{1}}. \tag{11.29}$$

However, since $\hat{\varphi}$ and $\hat{N}$ are conjugated variables, being $[\hat{\varphi}, \hat{N}] = i\hat{\mathbb{1}}$, choosing the basis of the eigenstates of $\hat{\varphi}$ we have the following correspondence:

$$\hat{\varphi} \to \varphi, \quad \text{and} \quad \hat{N} \to -i\frac{\partial}{\partial \varphi}, \tag{11.30}$$

and the Hamiltonian rewrites:

$$\hat{H} = 4E_c \left(-i\frac{\partial}{\partial \varphi} - N_g\right)^2 - E_J \cos\varphi. \tag{11.31}$$

The solutions of the differential equation:

$$\hat{H}\psi_m(\varphi) = E_m \psi_m(\varphi) \tag{11.32}$$

are given in terms of the Floquet-type solutions $\text{me}_\nu(q, x)$ as follows:

$$\psi_m(\varphi) = \frac{1}{\sqrt{2}} \text{me}_{-2[N_g - f(m, N_g)]} \left(-\frac{E_J}{2E_c}, \frac{\varphi}{2}\right), \tag{11.33}$$

with:

$$f(m, N_g) = \sum_{k=\pm 1} [\text{int}(2N_g + k/2) \bmod 2]$$

$$\times \left\{ \text{int}(N_g) - k(-1)^m \left[(m+1) \operatorname{div} 2 + m \bmod 2\right] \right\}, \tag{11.34}$$

11.2 The Josephson Junction and the SQUID

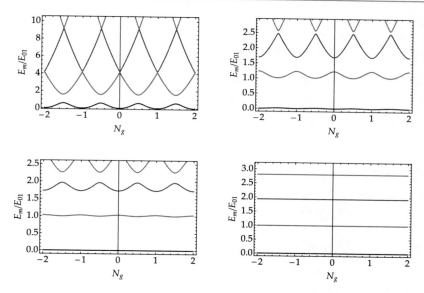

Fig. 11.2 Plots of E_m as a function of N_g (in each plot, from bottom to top $m = 0, 1, 2$ and 3) normalized with respect to $E_{01} \equiv \min_{N_g}(E_1 - E_0)$ for different values of the ratio E_J/E_c. (Top left) $E_J/E_c = 1.0$; (top right) $E_J/E_c = 5.0$; (bottom left) $E_J/E_c = 10.0$; (bottom right) $E_J/E_c = 50.0$. The zero point of energy is chosen as the bottom of the $m = 0$ level

where int(x) rounds to the integer closest to x, x mod y denotes the usual modulo operation, and x div y gives the integer quotient of x and y. The corresponding eigenvalues are:

$$E_m = E_c \, a_{-2[N_g - f(m, N_g)]}\left(-\frac{E_J}{2E_c}\right), \tag{11.35}$$

where $a_\nu(q)$ denotes Mathieu's characteristic value.

In Fig. 11.2 we report the behavior of E_m, $m = 0, 1, 2$, and 3, as a function of N_g and normalized with respect to the transition energy $E_{01} = E_1 - E_0$, which is the minimum energy separation between the levels E_1 and E_0, for different values of the ratio E_J/E_c.

As shown in Fig. 11.3 we can identify two regimes:

- the *charge regime* for $E_c \gg E_J$;
- and the so-called *transmon regime* for $E_c \ll E_J$.

The term "transmon" is an abbreviation of "transmission line shunted plasma oscillation qubit", as we will see in Sect. 11.5. In each of these regimes we can define a two level system which can be used as a qubit: according to the regime they are referred to as "charge qubit" and "transmon" qubit, respectively.

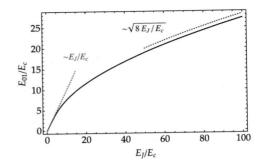

Fig. 11.3 Plot of E_{01}/E_c as a function of the ratio E_J/E_c: for $E_c \gg E_J$ (charge regime) we have $E_{01} \sim E_J$; for $E_c \ll E_J$ (transmon regime) we have $E_{01} \sim \sqrt{8 E_J E_c}$

11.3 The Charge Qubit

In the charge regime, $E_J \ll E_c$, our system can be seen as a Cooper pair box (CPB), that is sketched in Fig. 11.4. It consists in a superconducting electrode (the "island") in contact with a superconducting reservoir though a tunnel junction (the grey zone in figure, which corresponds to a Josephson junction or to the two junctions of the SQUID) with capacitance C_J. Excess Cooper pairs may tunnel onto the island in response to an electric field applied by means of the gate capacitance C_g and the voltage V_g.

In this case we have a well defined number N of tunnelling Cooper pairs and, thus, of excess of Cooper pairs, and a strongly fluctuating phase. Therefore we can express the Hamiltonian (11.28) as a function of the eigenstates $|N\rangle$ of $\hat{N}$, that is, $\hat{N}|N\rangle = N|N\rangle$, $N \in \mathbb{Z}$; we have:

$$\hat{H}_{\text{CPB}} = \sum_{N=-\infty}^{+\infty} \left[4E_c (N - N_g)^2 |N\rangle\langle N| \right.$$
$$\left. - \frac{1}{2} E_J (|N\rangle\langle N+1| + |N+1\rangle\langle N|) \right], \quad (11.36)$$

where the term $|N\rangle\langle N+1| + |N+1\rangle\langle N|$ describes the tunnelling through the junction of a single Cooper pair. It is now clear that E_J represents a measure of the coupling across the junction. It is worth noting that the states:

$$|\varphi\rangle = \frac{1}{\sqrt{2\pi}} \sum_{N=-\infty}^{+\infty} \exp(iN\varphi)|N\rangle \quad (11.37)$$

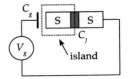

Fig. 11.4 Schematics of the CPB. The dashed box encloses the superconducting island

11.3 The Charge Qubit

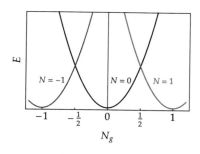 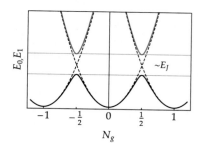

Fig. 11.5 Left plot: energy levels of the states $|N\rangle$ without interaction ($E_J = 0$): note the degeneracy at $N_g = (1+2N)/2$. Right plot: as $E_J \neq 0$ the degeneracy is broken, and, if $E_J \ll E_c$, we can identify two levels, E_0 (black) and E_1 (red), whose energy difference at $N_g = (1+2N)/2$ is $\sim E_J$

are eigenstates of the operator:

$$\hat{H}_{\text{tun}} = -\frac{1}{2} E_J \sum_{N=-\infty}^{+\infty} (|N\rangle\langle N+1| + |N+1\rangle\langle N|), \quad (11.38)$$

and $\hat{H}_{\text{tun}}|\varphi\rangle = -E_J \cos\varphi|\varphi\rangle$, that is we have the following expansion:

$$\cos\hat{\varphi} = \frac{1}{2} \sum_{N=-\infty}^{+\infty} (|N\rangle\langle N+1| + |N+1\rangle\langle N|). \quad (11.39)$$

If E_J is negligible, then $\hat{H}_{\text{CPB}}$ is just the sum of energies $4E_c(N-N_g)^2$ of the states $|N\rangle$ (see the left plot in Fig. 11.5): it is interesting to note that, for a particular choice of N_g, states with different number N may have the same energy (they are degenerate). In particular we can see that the two states $|N\rangle$ and $|N+1\rangle$ are degenerate if $N_g = (1+2N)/2$. As one may expect, the presence of the interaction, though weak but not negligible, breaks the degeneracy (see the right plot Fig. 11.5). In particular, an energy gap appears near degeneracy, which, for fixed N_g, allows us to identify two well-defined energy levels whose energy difference is E_J (see the top left plot in Fig. 11.2).

In fact, for a fixed N, and considering $N_g \approx (1+2N)/2$, we can assume that only the two states $|N\rangle$ and $|N+1\rangle$ are coupled by the interaction (this can be shown more rigorously by considering the interaction picture and the RWA). The corresponding two-level Hamiltonian can be written as:

$$\hat{H}_{\text{CPB}}(N_g, N) = 4E_c \left[(N-N_g)^2 |N\rangle\langle N| + (N+1-N_g)^2 |N+1\rangle\langle N+1| \right]$$
$$- \frac{1}{2} E_J \big(|N\rangle\langle N+1| + |N+1\rangle\langle N|\big), \quad (11.40)$$

$$= 4E_c \left[(N_g - N) - \frac{1}{2} \right] \overline{\sigma}_z^{(N)} - \frac{1}{2} E_J \overline{\sigma}_x^{(N)}$$
$$+ 2E_c \left[(N - N_g)^2 + (N - N_g + 1)^2 \right] |N+1\rangle\langle N+1|, \tag{11.41}$$

where we introduced:

$$\overline{\sigma}_z^{(N)} = |N\rangle\langle N| - |N+1\rangle\langle N+1|, \tag{11.42}$$

and:

$$\overline{\sigma}_x^{(N)} = |N\rangle\langle N+1| + |N+1\rangle\langle N|. \tag{11.43}$$

The eigenvalues of $\hat{H}_{\text{CPB}}(N_g, N)$ are:

$$E_{\text{CPB}}^{\pm}(N_g, N) = 2E_c \left[(N - N_g)^2 + (N - N_g + 1)^2 \right]$$
$$\pm \frac{1}{2} \sqrt{E_J^2 + 16E_c^2 [1 + 2(N - N_g)]^2}. \tag{11.44}$$

At degeneracy $N - N_g = -1/2$ and Eq. (11.40) rewrites (we neglect the constant term E_c):

$$\hat{H}_{\text{CPB}} \equiv \hat{H}_{\text{CPB}}(1/2, 0) = -\frac{1}{2} E_J \overline{\sigma}_x, \tag{11.45}$$

where $\overline{\sigma}_z = |0\rangle\langle 0| - |1\rangle\langle 1|$ and $\overline{\sigma}_x = |0\rangle\langle 1| + |1\rangle\langle 0|$. Since:

$$\hat{H}_{\text{CPB}} \rightarrow \begin{pmatrix} 0 & -\frac{1}{2} E_J \\ -\frac{1}{2} E_J & 0 \end{pmatrix}, \tag{11.46}$$

it is straightforward to find the two eigenvalues:

$$E_{\pm} = \pm \frac{1}{2} E_J, \quad \text{with} \quad E_+ - E_- = E_J, \tag{11.47}$$

and the corresponding eigenstates:

$$|e\rangle = \frac{|1\rangle - |0\rangle}{\sqrt{2}}, \quad |g\rangle = \frac{|1\rangle + |0\rangle}{\sqrt{2}}, \tag{11.48}$$

with $\hat{H}_{\text{CPB}}|e\rangle = E_+|e\rangle$ and $\hat{H}_{\text{CPB}}|g\rangle = E_-|g\rangle$, where we can see $|e\rangle$ as the "excited" state and $|g\rangle$ as the ground state of the system. Note that:

$$\bar{\sigma}_x = \underbrace{|g\rangle\langle g| - |e\rangle\langle e|}_{-\hat{\sigma}_z}, \quad \text{and} \quad \bar{\sigma}_z = \underbrace{|e\rangle\langle g| + |g\rangle\langle e|}_{\hat{\sigma}_x}, \quad (11.49)$$

where, as usual, $|e\rangle \to (1,0)^T$ and $|g\rangle \to (0,1)^T$. In the basis $\{|g\rangle, |e\rangle\}$, the Hamiltonian (11.45) simply reads (we neglect the constant term):

$$\hat{H}_{\text{CPB}} = \frac{\hbar \Omega}{2} \hat{\sigma}_z, \quad (11.50)$$

with

$$\Omega = \frac{E_J}{\hbar}, \quad (11.51)$$

that is the Hamiltonian of an *artificial atom* which can be used as a qubit.

As a matter of fact, the charge qubit is very sensible to the fluctuations of N_g and, thus, of the gate voltage V_g. This problem can be solved considering the so-called transmon regime (see Sect. 11.5).

11.4 Charge Qubit and Capacitive Coupling with a 1-D Resonator

A 1-D transmission line resonator consists of a full-wave section of a superconducting coplanar waveguide. If L_r and C_r are the effective inductance and capacitance of the resonator, respectively, then its characteristic frequency is $\omega_r = 1/\sqrt{L_r C_r}$ (typical values are $\omega_r \sim 10$ GHz). The quantum Hamiltonian of the resonator may be written as:

$$\hat{H}_r = \hbar \omega_r \left(\hat{a}^\dagger \hat{a} + \frac{1}{2} \right), \quad (11.52)$$

$\hat{a}$ and $\hat{a}^\dagger$ being the annihilation and creation operators, respectively, and $[\hat{a}, \hat{a}^\dagger] = \hat{\mathbb{I}}$. The 1-D resonator plays the role of the cavity of a cavity QED experiment and we have seen in Sect. 9.5.

As depicted in Fig. 11.6, a superconducting qubit, in the present case a CPB, is placed inside the 1-D resonator and it plays the role of the atom of the cavity QED setup. The system CPB+resonator is built in such a way that there is a maximum coupling between the qubit and resonator. As schematically shown in Fig. 11.6, the qubit couples with the mode 2 of the resonator (maxima at the center).

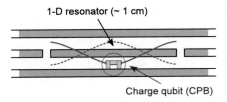

Fig. 11.6 Sketch of a typical configuration to implement circuit QED. A superconducting qubit (a CPB, in yellow) is built inside a 1-D transmission line resonator. The final configuration is such that there is a maximum coupling between the qubit and resonator (the rms voltages reaches the maxima at the center of the conductor, see the red lines)

The *free* Hamiltonian of the system reads:

$$\hat{H}_0 = \hbar\omega_r \left(\hat{a}^\dagger \hat{a} + \frac{1}{2}\right) + 4E_c(\hat{N} - N_g)^2 - E_J \cos\hat{\varphi}, \quad (11.53)$$

where the second and the third terms are the same as in Eq. (11.28).

The coupling between the resonator and the CPB is due to the presence of the quantum contribution to the voltage, which leads to the following substitution in Eq. (11.53):

$$N_g \to N_g + \hat{N}_r, \quad (11.54)$$

with:

$$\hat{N}_r = N_q (\hat{a}^\dagger + \hat{a}), \quad (11.55)$$

and:

$$N_q = \frac{C_g V_{\text{rms}}}{2e}, \quad (11.56)$$

where:

$$V_{\text{rms}} = \sqrt{\frac{\hbar\omega_r}{2C_r}} \quad (11.57)$$

is the rms voltage corresponding to the mode 2 of the resonator ($\omega_r \to \omega_r/2$) and C_g is the gate voltage. After the substitution we obtain the following Hamiltonian which describes also the coupling through the gate voltage (we neglect the constant term):

$$\hat{H} = \hbar\omega_r \hat{a}^\dagger \hat{a} + 4E_c \left[(\hat{N} - N_g) - N_q(\hat{a}^\dagger + \hat{a})\right]^2 - E_J \cos\hat{\varphi}, \quad (11.58)$$

11.5 The Transmon Qubit

In the following, we show that the presence of $\hat{H}_1$ is what we need to make the energy levels different enough in order to select a well defined two-level system.

Equation (11.66) represents the Hamiltonian of a nonlinear oscillator, therefore we can introduce the bosonic field annihilation, $\hat{b}$, and creation, $\hat{b}^\dagger$, operators, respectively, with $[\hat{b}, \hat{b}^\dagger] = \hat{\mathbb{I}}$, and put:

$$\hat{\varphi} = \left(\frac{2E_c}{E_J}\right)^{\frac{1}{4}} \left(\hat{b}^\dagger + \hat{b}\right) = 2\sqrt{\frac{E_c}{\hbar\omega_p}} \left(\hat{b}^\dagger + \hat{b}\right), \qquad (11.69)$$

$$\hat{N} = i\left(\frac{E_J}{32E_c}\right)^{\frac{1}{4}} \left(\hat{b}^\dagger - \hat{b}\right) = \frac{i}{4}\sqrt{\frac{\hbar\omega_p}{E_c}} \left(\hat{b}^\dagger - \hat{b}\right), \qquad (11.70)$$

where we introduced the *Josephson plasma frequency*:

$$\omega_p = \frac{\sqrt{8E_J E_c}}{\hbar}. \qquad (11.71)$$

It is easy to show that $[\hat{\varphi}, \hat{N}] = i\hat{\mathbb{I}}$ and that Eq. (11.66) becomes:

$$\hat{H}_{\text{tr}} = \underbrace{\hbar\omega_p \left(\hat{b}^\dagger \hat{b} + \frac{1}{2}\right)}_{\hat{H}_0} - \frac{1}{12} E_c \left(\hat{b}^\dagger + \hat{b}\right)^4, \qquad (11.72)$$

and $\hbar\omega_p = \sqrt{8E_J E_c}$.

Since $E_c \ll E_J$, to calculate the eigenvalues of Eq. (11.72) we can apply the first order perturbation theory. The unperturbed eigenvalues of $\hat{H}_{\text{tr}}$ are:

$$E_n^{(0)} = \hbar\omega_p \left(n + \frac{1}{2}\right), \qquad (11.73)$$

where $\hat{H}_0 |n\rangle = E_n^{(0)} |n\rangle$. The first order correction to $E_n^{(0)}$ is given by:

$$E_n^{(1)} = -\langle n | \left[\frac{1}{12} E_c \left(\hat{b}^\dagger + \hat{b}\right)^4\right] |n\rangle. \qquad (11.74)$$

and explicitly writes:

$$E_n^{(1)} = -\frac{1}{12} E_c \langle n | \left[12 \hat{b}^\dagger \hat{b} + 6(\hat{b}^\dagger)^2 \hat{b}^2 + 3 + \text{(terms s.t. } \langle n | \cdots | n \rangle = 0)\right] |n\rangle$$

$$= -E_c n - \frac{1}{2} E_c n(n-1) - \frac{1}{4} E_c. \qquad (11.75)$$

Therefore, neglecting the constant term, the perturbed energy levels are:

$$E_n = \left(\sqrt{8E_J E_c} - E_c\right) n - \frac{1}{2} E_c n(n-1). \tag{11.76}$$

It is worth noting that, due to the nonlinearity, the difference between adjacent levels is now dependent on n, namely:

$$\Delta E_{n,n+1} \equiv E_{n+1} - E_n = \left(\sqrt{8E_J E_c} - E_c\right) - E_c n. \tag{11.77}$$

In particular, we have:

$$\Delta E_{0,1} = \sqrt{8E_J E_c} - E_c, \tag{11.78}$$

$$\Delta E_{1,2} = \Delta E_{0,1} - E_c. \tag{11.79}$$

Since typical values of the involved quantities are $E_J/\hbar \approx 2$ GHz ad $E_c/\hbar \approx 400$ MHz (usually, $C_J \approx 10^{-12}$ F), it is possible to experimentally select only the transition between the levels E_0 and E_1, thus obtaining the so-called transmon qubit.

It is worth noting that the gain in charge-noise insensitivity, as E_J/E_c increases, leads also to a loss in anharmonicity. In order to reduce a many-level system to a qubit, that is a system with two well-defined levels, a sufficient anharmonicity is required. From the experimental point of view this sets a lower bound on the duration of control pulses to implement the quantum logic gates. However, it is possible to show that the energy ratio should satisfy $20 \lesssim E_J/E_c \ll 5 \cdot 10^4$, opening up a large range with exponentially decreased sensitivity to charge noise and yet sufficiently large anharmonicity for qubit operations.

Problems

11.1. ♣ Prove that if $\cos\hat{\varphi}|\varphi\rangle = \cos\varphi|\varphi\rangle$ with:

$$|\varphi\rangle = \frac{1}{\sqrt{2\pi}} \sum_{N=-\infty}^{+\infty} \exp(iN\varphi)|N\rangle,$$

then:

$$\frac{1}{2} \sum_{N=-\infty}^{+\infty} (|N\rangle\langle N+1| + |N+1\rangle\langle N|) = \cos\hat{\varphi}.$$

Further Readings

V. Bouchiat, D. Vion, P. Joyez, D. Esteve, M.H. Devoret, Quantum coherence with a single Cooper pair. Phys. Scr. **T76**, 165–170 (1998)

A. Blais, R.-S. Huang, A. Wallraff, S.M. Girvin, R.J. Schoelkopf, Cavity quantum electrodynamics for superconducting electrical circuits: an architecture for quantum computation. Phys. Rev. A **69**, 062320 (2004)

J. Koch, T.M. Yu, J. Gambetta, A.A. Houck, D.I. Schuster, J. Majer, A. Blais, M.H. Devoret, S.M. Girvin, R.J. Schoelkopf, Charge-insensitive qubit design derived from the Cooper pair box. Phys. Rev. A **76**, 042319 (2007)

M. Devoret, S. Girvin, R. Schoelkopf, Circuit-QED: How strong can the coupling between a Josephson junction atom and a transmission line resonator be? Ann. Phys. **16**, 767–779 (2007)

Quantum Computation and Adiabatic Evolution

12

Abstract

In the previous chapters we addressed quantum computation considering the quantum circuit model, where the information, encoded into qubits, is processed by means of quantum gates implementing a well defined algorithm. In this chapter we introduce a different approach to quantum computation, based on the adiabatic evolution. Here the problem is encoded into a given *problem Hamiltonian* and the solution is given by the *ground state* of the Hamiltonian itself. In order to find the solution, one starts form the ground state of an initial Hamiltonian which is then adiabatically transformed into the problem Hamiltonian. If the requirement of the so-called adiabatic theorem are satisfied, during the dynamics the system remains into the ground state of the *instantaneous* Hamiltonian and, thus, we finally end up in the ground state of the problem Hamiltonian.

12.1 Clauses and Instances of Satisfiability

In our context, a *clause* C is a boolean expression which can be *true* or *false* according to the values of the involved bits. For example, the two-bit clause:

$$x_1 \wedge x_2, \tag{12.1}$$

where $x_k \in \{0, 1\}$, is *true*, that is $x_1 \wedge x_2 = 1$, only if $x_1 = x_2 = 1$. We can also write the formal identity:

$$x_1 \wedge x_2 = x_1 x_2, \tag{12.2}$$

where at the r.h.s. we have the mathematical product of the bit values.

© The Author(s), under exclusive license to Springer Nature Switzerland AG 2025
S. Olivares, *A Student's Guide to Quantum Computing*, Lecture Notes in Physics 1038, https://doi.org/10.1007/978-3-031-83361-8_12

It is possible to associate a two-qubit Hamiltonian with the clause (12.1) whose ground state is the searched solution. In fact, recalling that $\hat{\sigma}_z|x\rangle = (-1)^x|x\rangle$, we have:

$$\frac{1}{2}\left(\hat{\mathbb{I}} - \hat{\sigma}_z\right)|x\rangle = x|x\rangle, \tag{12.3a}$$

$$\frac{1}{2}\left(\hat{\mathbb{I}} + \hat{\sigma}_z\right)|x\rangle = \overline{x}|x\rangle, \tag{12.3b}$$

and it is straightforward to check that the ground state $|x_1\rangle|x_2\rangle = |1\rangle|1\rangle$ of the Hamiltonian (for the sake of simplicity in this chapter we assume natural units, namely, we set $\hbar = 1$):

$$\hat{H}_{x_1 \wedge x_2} = \hat{\mathbb{I}} - \frac{1}{2}\left(\hat{\mathbb{I}} - \hat{\sigma}_z^{(1)}\right) \otimes \frac{1}{2}\left(\hat{\mathbb{I}} - \hat{\sigma}_z^{(2)}\right), \tag{12.4}$$

just encodes the solution of the clause (12.1). A similar result can be obtained for the clause:

$$x_1 \wedge \overline{x}_2 = x_1 \overline{x}_2, \tag{12.5}$$

and the corresponding Hamiltonian:

$$\hat{H}_{x_1 \wedge \overline{x}_2} = \hat{\mathbb{I}} - \frac{1}{2}\left(\hat{\mathbb{I}} - \hat{\sigma}_z^{(1)}\right) \otimes \frac{1}{2}\left(\hat{\mathbb{I}} + \hat{\sigma}_z^{(2)}\right), \tag{12.6}$$

whose ground state is $|1\rangle|0\rangle$.

Up to now we considered the AND operation. However, exploiting the logical identity $x_1 \vee x_2 = \text{NOT}(\overline{x}_1 \wedge \overline{x}_2)$, we can see that the Hamiltonian "solving" the clause $x_1 \vee x_2$ reads:

$$\hat{H}_{x_1 \vee x_2} = \frac{1}{2}\left(\hat{\mathbb{I}} + \hat{\sigma}_z^{(1)}\right) \otimes \frac{1}{2}\left(\hat{\mathbb{I}} + \hat{\sigma}_z^{(2)}\right), \tag{12.7}$$

that has the three degenerate ground states $|1\rangle|0\rangle$, $|0\rangle|1\rangle$ and $|1\rangle|1\rangle$.

As a last example, we consider a clause involving tree bits, an example of the so-called 3-SAT problem in which each clause involves just three bits, for example:

$$x_1 \vee x_2 \vee \overline{x}_3 \equiv \text{NOT}(\overline{x}_1 \wedge \overline{x}_2 \wedge x_3). \tag{12.8}$$

The reader can conclude that the Hamiltonian encoding the solution 001 in its ground state is:

$$\hat{H}_{x_1 \vee x_2 \vee \overline{x}_3} = \frac{1}{2}\left(\hat{\mathbb{I}} + \hat{\sigma}_z^{(1)}\right) \otimes \frac{1}{2}\left(\hat{\mathbb{I}} + \hat{\sigma}_z^{(2)}\right) \otimes \frac{1}{2}\left(\hat{\mathbb{I}} - \hat{\sigma}_z^{(3)}\right). \tag{12.9}$$

Inspecting the previous examples, it is easy finding the general rule to associate a suitable Hamiltonian with a logical clause. On the other hand, if writing that Hamiltonian is rather straightforward, retrieving the actual ground state cannot be easy at all ... For example, an n-bit instance of satisfiability is a logical expression:

$$C_1 \wedge C_2 \wedge \cdots \wedge C_M , \tag{12.10}$$

where C_k is a particular instance depending on the values of some subset of n bits. Finding a solution of a single clause could be simple but retrieving the solution of the whole instance (if it exists!) can be really difficult. Nevertheless, it is clear that we can associate a Hamiltonian $\hat{H}_k$, whose ground state is its specific solution, with any single clause C_k. Therefore, the total Hamiltonian:

$$\hat{H} = \hat{H}_1 + \hat{H}_2 + \cdots + \hat{H}_M \equiv \sum_{k=1}^{M} \hat{H}_k , \tag{12.11}$$

encodes in its ground state the solution of the instance (12.10) by construction. Notice that, for the sake of simplicity, we considered a scenario in which the ground state has zero energy.

Of course, now the problem is to find the ground state of the Hamiltonian (12.11) and, here, quantum mechanics can help.

12.2 The Adiabatic Theorem

In the previous section we have seen that we can associate a Hamiltonian $\hat{H}_p$ with a satisfiability problem in such a way that its ground state $|\Psi_p\rangle$ encodes the solution. In general, finding the ground state of $\hat{H}_p$ can be a hard problem. Nevertheless, we can solve it by applying the *adiabatic theorem*. In the following we will see the main ingredients of this theorem and its application to our purposes.

First of all we consider a Hamiltonian $\hat{H}_0$ whose ground state $|\Psi_0\rangle$ is easy to prepare and, then, we assume to have also a slowly-varying time-dependent *interpolating* Hamiltonian $\hat{H}(t)$ such that $\hat{H}(0) \equiv \hat{H}_0$ and $\hat{H}(T) \equiv \hat{H}_p$.

The time evolution of the state $|\psi(t)\rangle$ of the system is given as usual by the Schrödinger equation:

$$i\frac{d}{dt}|\psi(t)\rangle = \hat{H}(t)|\psi(t)\rangle . \tag{12.12}$$

If we define the parameter $s = t/T$, we can focus our analysis on the one-parameter family of Hamiltonians $\tilde{H}(s = t/T) \equiv \hat{H}(t)$ with $0 \leq s \leq 1$: the role of T is to control the rate at which $\hat{H}(t)$ changes, the longer T the slower the rate. Now we introduce the instantaneous eigenstates and eigenvalues of $\tilde{H}(s)$, namely:

$$\tilde{H}(s)|\phi_n(s)\rangle = E_n(s)|\phi_n(s)\rangle , \tag{12.13}$$

where

$$E_0(s) \leq E_1(s) \leq \cdots \leq E_{N-1}(s), \tag{12.14}$$

N being the dimension of the Hilbert space of the system. According to the adiabatic theorem, if $E_1(s) - E_0(s) > 0$, $\forall s \in [0, 1]$, we have:

$$\lim_{T \to \infty} |\langle \phi_0(1) | \psi(T) \rangle| = 1, \tag{12.15}$$

namely, if T is large enough, during the evolution the state $|\psi(t)\rangle$ of the system remains very close to the instantaneous ground state of the Hamiltonian $\hat{H}(t)$, $\forall t$. More in details, the adiabatic theorem states that if:

$$T \gg \frac{\Gamma_{\max}}{(\Delta E_{\min})^2}, \tag{12.16}$$

where:

$$\Gamma_{\max} = \max_{s \in [0,1]} \left| \left\langle \phi_1(s) \left| \frac{d\tilde{H}(s)}{ds} \right| \phi_0(s) \right\rangle \right|, \tag{12.17}$$

and:

$$\Delta E_{\min} = \min_{s \in [0,1]} [E_1(s) - E_0(s)], \tag{12.18}$$

then $|\langle \phi_0(1) | \psi(T) \rangle| \to 1$.

We are ready to apply the adiabatic theorem to the satisfiability problems.

12.3 Finding the Solutions Through the Adiabatic Evolution

The simplest interpolating, time-dependent Hamiltonian $\hat{H}(t)$ such that $\hat{H}(0) = \hat{H}_0$ and $\hat{H}(T) = \hat{H}_p$, i.e., the problem Hamiltonian, is:

$$\hat{H}(t) = \left(1 - \frac{t}{T}\right) \hat{H}_0 + \frac{t}{T} \hat{H}_p, \qquad t \in [0, T], \tag{12.19}$$

or, equivalently:

$$\tilde{H}(s) = (1 - s)\hat{H}_0 + s\hat{H}_p, \qquad s \in [0, 1]. \tag{12.20}$$

More in general, one can also use more sophisticated Hamiltonians substituting to the parameter s some other functions of the ratio t/T. It is clear that if we suitably choose T in order to satisfy the conditions of the adiabatic theorem, then

12.3 Finding the Solutions Through the Adiabatic Evolution

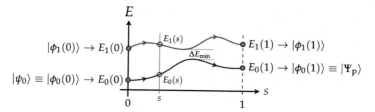

Fig. 12.1 In this plot we summarize the working principle of quantum computation assisted by the adiabatic evolution. In particular we plot two energy levels $E_0(s)$ and $E_1(s)$ as functions of $s = t/T$ assuming that they satisfy the adiabatic theorem: if the time T is large enough, the systems remains in its ground state during the whole evolution. See the text for details

we can let our system evolve from the initial ground state $|\psi_0\rangle \equiv |\phi_0(0)\rangle$ of the beginning Hamiltonian $\hat{H}_0$ and reach the ground state $|\Psi_p\rangle \equiv |\phi_0(1)\rangle$ of the problem Hamiltonian $\hat{H}_p$, as sketched in Fig. 12.1. If, however, the evolution is "too fast", then it is possible to obtain a final state at time $t = T$, i.e., $s = 1$, that is a linear combination of others energy eigenstates, thus finding, with a given probability, a wrong solution to the problem …

As a matter of fact, the beginning Hamiltonian should not be diagonal in the same basis of the problem one; otherwise the systems will always remain in the initial eigenstate, that, in general, is not the instantaneous ground state of the interpolating Hamiltonian, as we will see in the next section. Since, as we have mentioned in Sect. 12.1, at least in our cases the problem Hamiltonian in the presence of n qubits can be written as a function of the Pauli matrices $\hat{\sigma}_z^{(k)}$, $k = 1, \ldots, N$, a good choice for the starting Hamiltonian is:

$$\hat{H}_0 = \sum_{k=1}^{N} \hat{H}_0^{(k)}, \tag{12.21}$$

with:

$$\hat{H}_0^{(k)} = \frac{1}{2}\left(\hat{\mathbb{I}} - \hat{\sigma}_x^{(k)}\right). \tag{12.22}$$

It is worth noting that the corresponding ground state is (using the computational basis, namely, the eigenstates of $\hat{\sigma}_z^{(k)}$):

$$|\psi_0\rangle \equiv |\psi_0(s)\rangle = \frac{1}{2^{n/2}} \sum_{z=0}^{2^n-1} |z\rangle_n, \tag{12.23}$$

that is the balanced superposition of all the possible inputs. In the next section we will see a simple example based on a single qubit in order to see adiabatic computation at work.

12.4 One-qubit Example of Adiabatic Quantum Computation

Though this example is almost useless, we can use it to check the requirements of the adiabatic theorem and to follow the whole protocol analytically.

Here, we assume that our clause involves only one bit and it is simply:

$$C = z, \tag{12.24}$$

that is satisfied when $z = 1$ (of course!). The corresponding problem Hamiltonian reads (we still use dimensionless quantities):

$$\hat{H}_p = \frac{1}{2}\left(\hat{\mathbb{1}} + \hat{\sigma}_z\right), \tag{12.25}$$

and, as mentioned above, we consider the following beginning Hamiltonian:

$$\hat{H}_0 = \frac{1}{2}\left(\hat{\mathbb{1}} - \hat{\sigma}_x\right), \tag{12.26}$$

whose ground states is:

$$|\psi_0\rangle = \frac{1}{\sqrt{2}}\left(|0\rangle + |1\rangle\right). \tag{12.27}$$

The time-dependent Hamiltonian follows from Eq. (12.20) and, in the matrix representation, reads:

$$\tilde{H}(s) = \frac{1}{2}\begin{pmatrix} 1+s & s-1 \\ s-1 & 1-s \end{pmatrix}, \tag{12.28}$$

whose two eigenvalues:

$$E_0(s) = \frac{1}{2}\left(1 - \sqrt{2s^2 - 2s + 1}\right), \tag{12.29a}$$

$$E_1(s) = \frac{1}{2}\left(1 + \sqrt{2s^2 - 2s + 1}\right), \tag{12.29b}$$

are plotted in Fig. 12.2. We can see that the two levels are well-separated with $\Delta E_{\min} = 1/\sqrt{2}$: the adiabatic theorem can be applied and, after the evolution, the final state corresponds to the ground state of $\hat{H}_p$. Starting from Eq. (12.17) we also find $\Gamma_{\max} = 1/\sqrt{2}$ and, by using Eq. (12.16), we get $T \gg \sqrt{2}$ to achieve the adiabatic evolution.

12.4 One-qubit Example of Adiabatic Quantum Computation

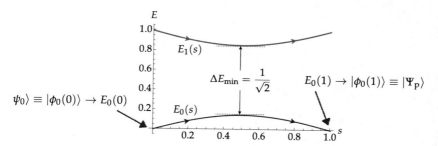

Fig. 12.2 Plot of the two eigenvalues of the Hamiltonian (12.28) as functions of s. We have explicitly highlighted the instantaneous eigenstates corresponding to the lowest eigenvalue at $s=0$ and $s=1$

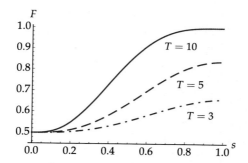

Fig. 12.3 Plot of the fidelity $F(s)$, with $s=t/T$, between the instantaneous eigenstate $|\phi_0(s)\rangle$, corresponding to the lowest eigenvalue of the Hamiltonian (12.28), and the evolved state $|\psi(T)\rangle$. The curves refer to different values of T: in the present case, the adiabatic theorem requires $T \gg \sqrt{2} \approx 1.41$. See the text for details

In order to assess the "success" of the computation, we introduce the *fidelity* between the final state $|\psi(T)\rangle$ and the instantaneous eigenstate $|\phi_0(s)\rangle$, namely:

$$F(s) = |\langle \phi_0(s)|\psi(T)\rangle|^2 . \qquad (12.30)$$

In Fig. 12.3 we plot $F(s)$ as a function of $s = t/T$ for different values of T: we can see that, as T increases, the fidelity at $s=1$ or, equivalently, at $t=T$, approaches 1, that is the state $|\psi(T)\rangle$ comes to coincide with the ground state of the Hamiltonian $\hat{H}(T) = \hat{H}_p$.

If we had chosen as starting Hamiltonian:

$$\hat{H}'_0 = \frac{1}{2}\left(\hat{\mathbb{1}} - \hat{\sigma}_z\right) , \qquad (12.31)$$

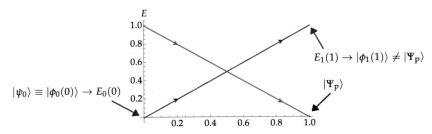

Fig. 12.4 Plot of the two eigenvalues of the Hamiltonian (12.32) as functions of s. Note that the lowest eigenvalue at $s = 0$ in transformed during the evolution in to the highest eigenvalue at $s = 1$. Since $\Delta E_{\min} = 0$, the adiabatic theorem cannot be applied

which clearly commutes with $\hat{H}_{\rm p}$, then we would have :

$$\tilde{H}'(s) = \begin{pmatrix} s & 0 \\ 0 & 1-s \end{pmatrix}, \qquad (12.32)$$

whose two eigenvalues become equal at $s = 1/2$ (see Fig. 12.4). In this case at the end of the evolution the system has passed from the ground state of $\hat{H}_0$ to the excited state of $\hat{H}_{\rm p}$: remarkably, they formally coincide, since the initial state is an eigenvector of $\tilde{H}'(s)$ and, thus, it is left unchanged during the whole evolution, up to a global phase.

It is interesting to note that to remove the level degeneracy it is enough to add a perturbation to the starting Hamiltonian, in such a way that the resulting one does no longer commute with $H_{\rm p}$. In particular, if we consider:

$$\hat{H}'_0 = \frac{1}{2}\left(\hat{\mathbb{1}} - \hat{\sigma}_z\right) + \varepsilon\,\hat{\sigma}_x, \qquad (12.33)$$

with $0 < \varepsilon \ll 1$, we have:

$$\tilde{H}'(s) = \begin{pmatrix} s & \varepsilon(1-s) \\ \varepsilon(1-s) & 1-s \end{pmatrix}, \qquad (12.34)$$

and the two corresponding eigenvalues read:

$$E_0(s) = \frac{1}{2}\left[1 - \sqrt{(2s-1)^2 + 4\varepsilon^2(1-s)^2}\right], \qquad (12.35a)$$

$$E_1(s) = \frac{1}{2}\left[1 + \sqrt{(2s-1)^2 + 4\varepsilon^2(1-s)^2}\right], \qquad (12.35b)$$

12.5 Factorization with Adiabatic Evolution

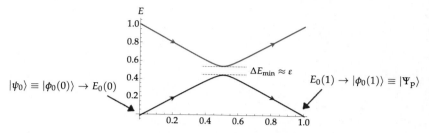

Fig. 12.5 Plot of the two eigenvalues of the Hamiltonian (12.32) as functions of s. We have explicitly highlighted the instantaneous eigenstates corresponding to the lowest eigenvalue at $s = 0$ and $s = 1$

which are plotted in Fig. 12.5. Now, the degeneracy at $s = 1/2$ is broken and, by suitably choosing T, the adiabatic theorem can be applied. Note that, in this particular case we have:

$$\Delta E_{\min} = \frac{\varepsilon}{\sqrt{1+\varepsilon^2}} \approx \varepsilon, \qquad (12.36)$$

occurring at:

$$s_{\min} = \frac{1+2\varepsilon^2}{2(1+\varepsilon^2)} \approx \frac{1}{2}. \qquad (12.37)$$

12.5 Factorization with Adiabatic Evolution

In Sect. 5.3 we discussed the Shor's algorithm for the factorization of large integer numbers. That algorithm is based on the circuit model of quantum computation. However, it has been proved that the adiabatic model of quantum computation and the circuit model are equivalent. In the following we will describe a factoring algorithm based on the adiabatic evolution that has been also experimentally implemented using NMR.

The problem we address is to find the *two integer prime factors*, p and q, of an integer number N encoded into $L = \lceil \log_2 N \rceil$ bits. It is worth noting that we are assuming from the beginning that there are *only two factors*. In order to solve our problem by adiabatic evolution, we should turn it into a problem of optimization, that will allow us to define the beginning and the problem Hamiltonians.

First of all, we introduce the function:

$$f(x, y) = (N - xy)^2, \qquad (12.38)$$

and it is clear that $f(x, y) \geq 0$ and $f(x, y) \geq 0$ only if $xy = pq \equiv N$. The problem Hamiltonian is introduced assuming that $f(x, y)$ are its eigenvalues, namely:

$$\hat{H}_p = \sum_{x,y} f(x, y) |x, y\rangle\langle x, y|, \qquad (12.39)$$

where $|x, y\rangle = |x\rangle|y\rangle$, while $|x\rangle = |x\rangle_X$ and $|y\rangle = |y\rangle_Y$ encode the two factors x and y, respectively, X and Y being the numbers of qubits used.

At this point we make two reasonable assumptions to simplify the problem:

1. N is odd, otherwise one factor is the number 2. Therefore, we know that x and y should be odd and we save one bit, since if $|x\rangle = |x_{X-1}\rangle \cdots |x_0\rangle$ with binary expansion $x = \sum_{k=0}^{X-1} 2^k x_k$, we have $x_0 \equiv 1$.
2. $x < y$, that is we can take $3 \leq x \leq \sqrt{N}$ and $\sqrt{N} \leq y \leq N/3$.

These assumptions allow us evaluating the effective number of qubits n_x and n_y needed to encode the factors, namely:

$$n_x \leq \left\lfloor \frac{L+1}{2} \right\rfloor \quad \text{and} \quad n_y \leq L - 1. \qquad (12.40)$$

We can conclude that the total number $n = n_x + n_y$ of qubits scales as $n \sim O(\frac{3}{2}L)$. Here we recall that, in the case of the Shor's algorithm, the number of needed bits is $n = 2L + 1 + \lceil \log[2 + (2\varepsilon)^{-1}] \rceil$, ε being the failure probability (see Sect. 5.3).

We now turn the attention to the problem Hamiltonian. Since the two factors are odd, we actually need $(n_x - 1) + (n_y - 1) = n - 2$ qubits (we exclude from the count the last qubit of each factor which is always 1). Following the method introduced in Sect. 12.1, the Hamiltonian can be written as:

$$\hat{H}_p = \left[N\hat{\mathbb{I}} - \left(2^{n_x - 1} \frac{\hat{\mathbb{I}} - \hat{\sigma}_z^{(1)}}{2} + \cdots 2^1 \frac{\hat{\mathbb{I}} - \hat{\sigma}_z^{(n_x - 1)}}{2} + 2^0 \hat{\mathbb{I}} \right) \right.$$
$$\left. \otimes \left(2^{n_y - 1} \frac{\hat{\mathbb{I}} - \hat{\sigma}_z^{(n_x)}}{2} + \cdots 2^1 \frac{\hat{\mathbb{I}} - \hat{\sigma}_z^{(n-2)}}{2} + 2^0 \hat{\mathbb{I}} \right) \right]^2, \qquad (12.41)$$

which acts on the states:

$$|x\rangle_{n_x - 1} |y\rangle_{n_y - 1} = |x_{n_x - 1}\rangle \cdots |x_1\rangle |y_{n_y - 1}\rangle \cdots |y_1\rangle, \qquad (12.42)$$

$$= \underbrace{|z_1\rangle \cdots |z_{n_x - 1}\rangle}_{n_x - 1} \underbrace{|z_{n_x}\rangle \cdots |z_{n-2}\rangle}_{n_y - 1} \equiv |z\rangle_{n-2}. \qquad (12.43)$$

12.5 Factorization with Adiabatic Evolution

Note that:

$$\sum_{k=1}^{n_x-1} 2^k x_k = 2 \sum_{k=0}^{n_x-2} 2^k x_{k+1} = 2x, \tag{12.44}$$

and

$$\sum_{k=1}^{n_y-1} 2^k y_k = 2 \sum_{k=0}^{n_y-2} 2^k y_{k+1} = 2y, \tag{12.45}$$

where x and y are encoded into $n_x - 1$ and $n_y - 1$ bits, respectively. Therefore we have:

$$\hat{H}_p |z\rangle_{n-2} = \left[N - \left(2^{n_x-1} z_1 + \cdots + 2^1 z_{n_x-1} + 1 \right) \right.$$

$$\left. \times \left(2^{n_y-1} z_{n_x} + \cdots + 2^1 z_{n-2} + 1 \right) \right]^2 |z\rangle_{n-2}, \tag{12.46}$$

$$= [N - (2x+1)(2y+1)]^2 |z\rangle_{n-2}. \tag{12.47}$$

The lowest eigenvalue of H_p is 0 corresponding to the state $|p'\rangle_{n_x-1} |q'\rangle_{n_y-1}$, which encodes the two numbers $p = 2p' + 1$ and $q = 2q' + 1$, that is the solution of our problem.

As beginning Hamiltonian we can choose, for instance:

$$\hat{H}_0 = \gamma \sum_{k=1}^{n-2} \hat{\sigma}_x^{(k)}, \tag{12.48}$$

with ground state:

$$|\psi_0\rangle_{n-2} = \frac{1}{2^{(n-2)/2}} \sum_{z=1}^{2^{n-2}-1} (-1)^{\Pi(z)} |z\rangle_{n-2}, \tag{12.49}$$

where $\Pi(z)$ is the parity of the number z, that is the number of 1s in the binary representation modulo 2. Interestingly, the Hamiltonian (12.48) describes a system, in which all the spins interact with the same magnetic field oriented along the x direction, γ being the coupling strength.

As a matter of fact, in order to eventually apply the adiabatic theorem, one should verify whether the conditions underlaying it are satisfied and this requires to choose a particular N and to study the corresponding problem Hamiltonian. The adiabatic protocol has been experimentally applied to the factorization of $N = 21$

by using NMR techniques and the interested reader can find further details about the experiment in suggested further readings.

Further Readings

E. Farhi, J. Goldstone, S. Gutmann, M. Sipser, *Quantum Computation by Adiabatic Evolution.* MIT-CTP-2936 (2000). arXiv:quant-ph/0001106v1

E. Messiah, *Quantum Mechanics*, vol. II, chap. XVII, sect. 13 (Holland Publishing Company, Amsterdam, 1965)

X. Peng, Z. Liao, N. Xu, G. Qin, X. Zhou, D. Suter, J. Du, A quantum adiabatic algorithm for factorization and its experimental implementation. Phys. Rev. Lett. **101**, 220405 (2008)

Interaction Picture

Given the Hamiltonian:

$$\hat{H} = \hat{H}_0 + \hat{H}_{\text{int}}, \tag{A.1}$$

$\hat{H}_0$ and $\hat{H}_{\text{int}}$ being the free and interaction Hamiltonian, respectively, it is sometime useful to use the so-called *interaction picture*. If $|\psi_t\rangle$ represents the state of the system at the time t, its evolution is governed by the Schrödinger equation:

$$i\hbar \frac{\partial}{\partial t} |\psi_t\rangle = \hat{H} |\psi_t\rangle. \tag{A.2}$$

Now, we apply the following unitary transformation:

$$|\psi_t\rangle \to |\psi'_t\rangle = \hat{U}_0(t)^\dagger |\psi_t\rangle \quad \Rightarrow \quad |\psi_t\rangle = \hat{U}_0(t) |\psi'_t\rangle \tag{A.3}$$

where $\hat{U}_0(t) = \exp(-i\hat{H}_0 t/\hbar)$. Substituting into the Schrödinger equation we have:

$$i\hbar \frac{\partial}{\partial t} \left[\hat{U}_0(t) |\psi'_t\rangle \right] = \left(\hat{H}_0 + \hat{H}_{\text{int}} \right) \hat{U}_0(t) |\psi'_t\rangle, \tag{A.4}$$

$$\hat{H}_0 \hat{U}_0(t) |\psi'_t\rangle + i\hbar \hat{U}_0(t) \frac{\partial}{\partial t} |\psi'_t\rangle = \left(\hat{H}_0 + \hat{H}_{\text{int}} \right) \hat{U}_0(t) |\psi'_t\rangle, \tag{A.5}$$

and, after some algebra and applying $\hat{U}_0^\dagger(t)$ to both sides, we obtain:

$$i\hbar \frac{\partial}{\partial t} |\psi'_t\rangle = \hat{H}'_{\text{int}}(t) |\psi'_t\rangle, \tag{A.6}$$

where we introduced $\hat{H}'_{\text{int}}(t) = \hat{U}_0^\dagger(t)\hat{H}_{\text{int}}\hat{U}_0(t)$. Therefore, by using the interaction picture with respect to the free Hamiltonian,[1] one can focus only on the (transformed) interaction Hamiltonian.

---

[1] More in general, one can perform the interaction picture considering a different Hamiltonian which, in the case under investigation, allows to simplify the description of the system.

The Fabry–Perot Cavity

B

The main interaction between light and atoms in quantum electrodynamics (QED) is the dipolar interaction. On the one hand, the dipole moment is fixed by the nature of the atom: usually experimentalists use the Rydberg states (that is states with very high principal quantum number n in order to obtain a high electric dipole moment) of alkali atoms, such as Rb atoms. On the other hand, one can realize a very large electric field in a narrow band of frequencies and in a small volume of space by means of a Fabry–Perot cavity.

A Fabry–Perot cavity consists of two semi-reflecting mirrors with reflectivity R_1 and R_2, respectively. In order to find the field inside the cavity, we consider what happens when two classical fields $E_a^{(\text{in})}$ and $E_b^{(\text{in})}$ are mixed at a semi-reflecting mirror with reflectivity R (see Fig. B.1). If we denote with $E_a^{(\text{out})}$ and $E_b^{(\text{out})}$ the output field, we have the following linear transformation:

$$\begin{pmatrix} E_a^{(\text{out})} \\ E_b^{(\text{out})} \end{pmatrix} = \begin{pmatrix} \sqrt{R} & \sqrt{1-R} \\ \sqrt{1-R} & -\sqrt{R} \end{pmatrix} \begin{pmatrix} E_a^{(\text{in})} \\ E_b^{(\text{in})} \end{pmatrix} \qquad (\text{B.1})$$

that is:

$$E_a^{(\text{out})} = \sqrt{R}\, E_a^{(\text{in})} + \sqrt{1-R}\, E_b^{(\text{in})}, \qquad (\text{B.2})$$

$$E_b^{(\text{out})} = -\sqrt{R}\, E_b^{(\text{in})} + \sqrt{1-R}\, E_a^{(\text{in})}. \qquad (\text{B.3})$$

The scheme of the Fabry–Perot cavity is sketched in Fig. B.2: two mirrors with reflectivity R_1 and R_2, respectively, are placed at a distance L. The cavity is pumped with an input field $E^{(\text{in})}$ of frequency ω, which impinges on the first mirror. The transmitted part undergoes multiple reflections between the two mirrors leading to an overall forward and backward field inside the cavity, $E_{\text{fwd}}^{(\text{cav})}$ and $E_{\text{bwd}}^{(\text{cav})}$, respectively, an overall transmitted field $E^{(\text{out})}$ and an overall reflected field $E^{(\text{rfl})}$,

Fig. B.1 Input and output fields at a semi-reflecting mirror with reflectivity R

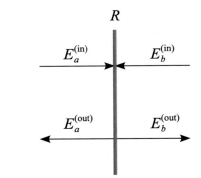

Fig. B.2 Scheme of Fabry–Perot cavity. See the text for details

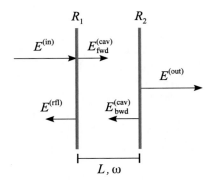

as depicted in Fig. B.2. If we define $\phi = 2L\omega/c$, then we have:

$$E_{\text{fwd}}^{(\text{cav})} = \frac{\sqrt{1-R_1}}{1 + e^{i\phi}\sqrt{R_1 R_2}} E^{(\text{in})}, \tag{B.4a}$$

$$E_{\text{bwd}}^{(\text{cav})} = e^{i\phi/2}\sqrt{R_2}\, E_{\text{fwd}}^{(\text{cav})}, \tag{B.4b}$$

and

$$E^{(\text{out})} = e^{i\phi/2}\sqrt{1-R_2}\, E_{\text{fwd}}^{(\text{cav})}, \tag{B.5a}$$

$$E^{(\text{rfl})} = e^{i\phi}\sqrt{(1-R_1)R_2}\, E_{\text{fwd}}^{(\text{cav})} + \sqrt{R_1}\, E^{(\text{in})}. \tag{B.5b}$$

In particular, if we assume $R_1 = R_2 = R$ and choose L such that $\phi = (2m+1)\pi$ (field-cavity resonance condition), $m \in \mathbb{N}$, we obtain:

$$E_{\text{fwd}}^{(\text{cav})} = \frac{E^{(\text{in})}}{\sqrt{1-R}}, \tag{B.6a}$$

B The Fabry–Perot Cavity

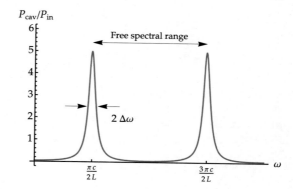

Fig. B.3 Ratio between the input power and the power of the field inside the cavity as a function of the input field frequency ω. We set $R_1 = R_2 = 0.8$. See the text for details

$$E^{(\text{cav})}_{\text{bwd}} = i \, \frac{\sqrt{R}}{\sqrt{1-R}} \, E^{(\text{in})}, \tag{B.6b}$$

$$E^{(\text{out})} = i \, E^{(\text{in})}, \quad E^{(\text{rfl})} = 0. \tag{B.6c}$$

A quantity usually considered to investigate the behavior of the cavity is the ratio between the input field power and the forward cavity field power, namely:

$$\frac{P_{\text{cav}}}{P_{\text{in}}} = \left| \frac{E^{(\text{cav})}_{\text{fwd}}}{E^{(\text{in})}} \right|^2 = \frac{1 - R_1}{1 + R_1 R_2 + 2\sqrt{R_1 R_2} \, \cos \phi}. \tag{B.7}$$

In Fig. B.3 we plot $P_{\text{cav}}/P_{\text{in}}$ as a function of the input field frequency: it is clear that near resonance we have a high field amplitude inside the cavity.

In order to better understand the behavior of the ratio defined in Eq. (B.7) we introduce $\delta = \phi - \pi$, i.e., the resonance is obtained for $\delta = 0$, and we consider the limit $\delta \ll 1$. We get the following expression for Eq. (B.7):

$$\frac{P_{\text{cav}}}{P_{\text{in}}} = \frac{1 - R_1}{(1 - \sqrt{R_1 R_2})^2} \frac{\Delta^2(R_1, R_2)}{\delta^2 + \Delta^2(R_1, R_2)} \tag{B.8}$$

that is a Lorentzian function where the half-width at half-maximum (HWHM) is:

$$\Delta^2(R_1, R_2) = \frac{(1 - \sqrt{R_1 R_2})^2}{\sqrt{R_1 R_2}}, \tag{B.9}$$

which, assuming $R_1 = R_2 = R$, reduces to:

$$\Delta(R) = \frac{1 - R}{\sqrt{R}}, \tag{B.10}$$

corresponding to a spectral bandwidth HWHM:

$$\Delta\omega = \frac{c}{2L}\frac{1-R}{\sqrt{R}}. \tag{B.11}$$

Another important quantity is the cavity finesse, that is the ratio between the free spectral range, and the full-width half-maximum (FWHM) of Eq. (B.7) at resonance. In the present case the free spectral range is $2\pi c/(2L)$ (see Fig. B.3), while the FWHM is $2\Delta\omega$, thus the cavity finesse is given by:

$$\mathcal{F} = \frac{2\pi c}{2L}\frac{1}{2\Delta\omega} = \pi\frac{\sqrt{R}}{1-R}. \tag{B.12}$$

The reader can obtain a quantitative analysis of the cavities involved in typical cavity QED experiments considering that $R \approx 1$ and $L \sim 1$ cm: this is why we have a very high field amplitude inside the cavity in the microwave domain, and, remarkably, microwaves are the characteristic transition frequencies of the Rydberg states involved in these experiments.

We now focus the attention on plain waves and assume that the axis of the cavity is aligned with the z-axis of a reference frame, where the mirrors are placed at $z = 0$ and $z = L$, respectively. Inside the cavity we have two counterpropagating waves [here we also assume to be at resonance and use Eq. (B.6)]:

$$E_{\text{fwd}}^{(\text{cav})}(z,\omega) = \frac{E^{(\text{in})}}{\sqrt{1-R}}\cos(kz-\omega t), \tag{B.13a}$$

$$E_{\text{bwd}}^{(\text{cav})}(z,\omega) = -\frac{E^{(\text{in})}\sqrt{R}}{\sqrt{1-R}}\sin(kz+\omega t), \tag{B.13b}$$

therefore, inside the cavity we have the following wave:

$$E_{\text{cav}}(z,\omega) = \frac{E^{(\text{in})}}{\sqrt{1-R}}\left[\cos(kz-\omega t) - \sqrt{R}\sin(kz+\omega t)\right]. \tag{B.14}$$

If we now perform the time average of the intensity of the field inside the cavity, we find:

$$\left\langle |E_{\text{cav}}(z)|^2 \right\rangle \equiv \frac{\omega}{2\pi}\int_0^{2\pi/\omega} |E_{\text{cav}}(z,\omega)|^2 \, dt \tag{B.15}$$

$$= \frac{1+R-2\sqrt{R}\sin(2kz)}{2(1-R)}\left|E^{(\text{in})}\right|^2 \tag{B.16}$$

$$= \frac{1+R-2\sqrt{R}\sin\left[(2m+1)\pi\frac{z}{L}\right]}{2(1-R)}\left|E^{(\text{in})}\right|^2, \tag{B.17}$$

B The Fabry–Perot Cavity

where, in the last equality, we used the resonance condition for the wave vector $k = \omega/c$, namely $k = (2m + 1)\pi/(2L)$. In the case of optical frequencies $m \approx 10^5$ and if we consider the average over the z direction we find:

$$\left\langle |E_{\text{cav}}|^2 \right\rangle \approx \frac{1+R}{2(1-R)} \left|E^{(\text{in})}\right|^2. \tag{B.18}$$

Solutions

Here we report the solutions of the problems marked with the symbol "♣".

Problems of Chap. 1

1.5 One possibility is to perform the whole calculation exploiting the expansion over the Pauli matrices:

$$\mathbf{C}_{hk} = \frac{1}{2}\left(\hat{\mathbb{I}} + \hat{\sigma}_x^{(k)} + \hat{\sigma}_z^{(h)} - \hat{\sigma}_z^{(h)}\hat{\sigma}_x^{(k)}\right).$$

However, we can also show that $\mathbf{C}_{hk}\mathbf{C}_{kh}\mathbf{C}_{hk}$ acts as $\mathbf{S}_{hk}$ on the computational basis. In this case it is straightforward to see that $|0_h\rangle|0_k\rangle$ is left unchanged as for $\mathbf{S}_{hk}$, whereas for the other basis elements we have:

$$\mathbf{C}_{hk}\mathbf{C}_{kh}\mathbf{C}_{hk}|0_h\rangle|1_k\rangle = \mathbf{C}_{hk}\mathbf{C}_{kh}|0_h\rangle|1_k\rangle$$
$$= \mathbf{C}_{hk}|1_h\rangle|1_k\rangle = |1_h\rangle|0_k\rangle \equiv \mathbf{S}_{hk}|0_h\rangle|1_k\rangle;$$

$$\mathbf{C}_{hk}\mathbf{C}_{kh}\mathbf{C}_{hk}|1_h\rangle|0_k\rangle = \mathbf{C}_{hk}\mathbf{C}_{kh}|1_h\rangle|1_k\rangle$$
$$= \mathbf{C}_{hk}|0_h\rangle|1_k\rangle = |0_h\rangle|1_k\rangle \equiv \mathbf{S}_{hk}|1_h\rangle|0_k\rangle;$$

$$\mathbf{C}_{hk}\mathbf{C}_{kh}\mathbf{C}_{hk}|1_h\rangle|1_k\rangle = \mathbf{C}_{hk}\mathbf{C}_{kh}|1_h\rangle|0_k\rangle$$
$$= \mathbf{C}_{hk}|1_h\rangle|0_k\rangle = |1_h\rangle|1_k\rangle \equiv \mathbf{S}_{hk}|1_h\rangle|1_k\rangle.$$

1.6 The Hadamard transformation can be written as:

$$\mathbf{H} = \frac{1}{\sqrt{2}}(\mathbf{X} + \mathbf{Z}),$$

therefore we have:

$$\mathbf{HX} = \frac{1}{\sqrt{2}} (\mathbf{XX} + \mathbf{ZX}) = \frac{1}{\sqrt{2}} \left(\hat{\mathbb{I}} + i\mathbf{Y} \right),$$

and

$$(\mathbf{HX})\mathbf{H} = \frac{1}{2} (\mathbf{X} + \mathbf{Z} + i\mathbf{YX} + i\mathbf{YZ}) = \frac{1}{2} (\mathbf{X} + \mathbf{Z} + \mathbf{Z} - \mathbf{X}) = \mathbf{Z},$$

where we used $\mathbf{ZX} = i\mathbf{Y}$, $\mathbf{YX} = -i\mathbf{Z}$ and $\mathbf{YZ} = i\mathbf{X}$. Analogously, one can prove that $\mathbf{HZH} = \mathbf{X}$.

1.7 Since:

$$\mathbf{C}_{hk} = \frac{1}{2} \left(\hat{\mathbb{I}} + \hat{\sigma}_x^{(k)} + \hat{\sigma}_z^{(h)} - \hat{\sigma}_z^{(h)} \hat{\sigma}_x^{(k)} \right)$$

we have:

$$\mathbf{H}_h \mathbf{H}_k \mathbf{C}_{hk} \mathbf{H}_h \mathbf{H}_k = \frac{1}{2} \left(\hat{\mathbb{I}} + \mathbf{H}_k \hat{\sigma}_x^{(k)} \mathbf{H}_k + \mathbf{H}_h \hat{\sigma}_z^{(h)} \mathbf{H}_h - \mathbf{H}_h \hat{\sigma}_z^{(h)} \mathbf{H}_h \mathbf{H}_k \hat{\sigma}_x^{(k)} \mathbf{H}_k \right).$$

Applying the result of problem 1.6 we can write:

$$\mathbf{H}_h \mathbf{H}_k \mathbf{C}_{hk} \mathbf{H}_h \mathbf{H}_k = \frac{1}{2} \left(\hat{\mathbb{I}} + \hat{\sigma}_z^{(k)} + \hat{\sigma}_x^{(h)} - \hat{\sigma}_x^{(h)} \hat{\sigma}_z^{(k)} \right) \equiv \mathbf{C}_{kh}.$$

Problems of Chap. 2

2.3 The 2×2 matrix associated with the considered Hamiltonian is (without loss of generality we can assume the coupling constant $\gamma \in \mathbb{R}$):

$$\hat{H} \to \begin{pmatrix} E_0 & g \\ g & E_1 \end{pmatrix}$$

where $E_k = \hbar \omega_k$, $k = 0, 1$, and $g = \hbar \gamma$. The eigenvalues are:

$$E_\pm = \frac{(E_0 + E_1) \pm \sqrt{(\Delta E)^2 + 4g^2}}{2},$$

with $\Delta E = E_1 - E_0$, and the corresponding eigenvectors $|\psi_\pm\rangle$, $\hat{H}|\psi_\pm\rangle = E_\pm |\psi_\pm\rangle$, can be written as:

$$|\psi_\pm\rangle = c_{0,\pm} |0\rangle + c_{1,\pm} |1\rangle,$$

where:

$$c_{0,\pm} = \frac{g}{\sqrt{(E_\pm - E_0)^2 + g^2}},$$

$$c_{1,\pm} = \frac{E_\pm - E_0}{\sqrt{(E_\pm - E_0)^2 + g^2}}.$$

In order to calculate the time evolution of the states $|0\rangle$ and $|1\rangle$, we rewrite them as functions of $|\psi_\pm\rangle$, namely:

$$|0\rangle = \frac{(E_+ - E_0)\sqrt{(E_- - E_0)^2 + g^2}\,|\psi_-\rangle - (E_- - E_0)\sqrt{(E_+ - E_0)^2 + g^2}\,|\psi_+\rangle}{g(E_+ - E_-)},$$

$$|1\rangle = \frac{\sqrt{(E_+ - E_0)^2 + g^2}\,|\psi_+\rangle - \sqrt{(E_- - E_0)^2 + g^2}\,|\psi_-\rangle}{E_+ - E_-},$$

or, in a more compact form:

$$|0\rangle = a_+ |\psi_+\rangle + a_- |\psi_-\rangle \quad \text{and} \quad |1\rangle = b_+ |\psi_+\rangle + b_- |\psi_-\rangle,$$

where:

$$a_\pm = \pm \frac{(E_\pm - E_0)\sqrt{(E_\pm - E_0)^2 + g^2}}{g(E_+ - E_-)} \quad \text{and} \quad b_\pm = \pm \frac{g\,a_\pm}{E_\pm - E_0}.$$

Thereafter, we can write the time evolution of the initial state $|1\rangle$ as follows:

$$|1\rangle = e^{-i\omega_+ t}\, b_+ |\psi_+\rangle + e^{-i\omega_- t}\, b_- |\psi_-\rangle,$$

with $\hbar \omega_\pm = E_\pm$.

2.4 First of all we note that:

$$(\boldsymbol{n} \cdot \boldsymbol{\sigma})^2 = \mathbb{1},$$

due to the properties of the Pauli matrices and since $|\boldsymbol{n}| = 1$. Therefore, we can divide the expansion of the exponential as follows:

$$\exp(i\gamma\, \boldsymbol{n} \cdot \boldsymbol{\sigma}) = \underbrace{\sum_{k=0}^{\infty} \frac{(-1)^k}{(2k)!} (\gamma)^{2k}}_{\cos\gamma} \mathbb{1} + i \underbrace{\sum_{k=0}^{\infty} \frac{(-1)^k}{(2k+1)!} (\gamma)^{2k+1}}_{\sin\gamma} (\boldsymbol{n} \cdot \boldsymbol{\sigma}).$$

Problems of Chap. 3

3.1 The overall state of the four qubits reads:

$$|\psi_{ABCD}\rangle = \frac{1}{2}\Big(|0_A\rangle|0_B\rangle|0_C\rangle|0_D\rangle + |0_A\rangle|0_B\rangle|1_C\rangle|1_D\rangle$$
$$+ |1_A\rangle|1_B\rangle|0_C\rangle|0_D\rangle + |1_A\rangle|1_B\rangle|1_C\rangle|1_D\rangle\Big).$$

Since we perform the Bell measurement on the subsystem of qubits B and C, its is useful to introduce the corresponding Bell basis:

$$\left|\Psi_{BC}^{(\pm)}\right\rangle = \frac{|0_B\rangle|1_C\rangle \pm |1_B\rangle|0_C\rangle}{\sqrt{2}},$$

$$\left|\Phi_{BC}^{(\pm)}\right\rangle = \frac{|0_B\rangle|0_C\rangle \pm |1_B\rangle|1_C\rangle}{\sqrt{2}},$$

and, in turn, we have:

$$|0_B\rangle|0_C\rangle = \frac{1}{\sqrt{2}}\left(\left|\Phi_{BC}^{(+)}\right\rangle + \left|\Phi_{BC}^{(-)}\right\rangle\right),$$

$$|0_B\rangle|1_C\rangle = \frac{1}{\sqrt{2}}\left(\left|\Psi_{BC}^{(+)}\right\rangle + \left|\Psi_{BC}^{(-)}\right\rangle\right),$$

$$|1_B\rangle|0_C\rangle = \frac{1}{\sqrt{2}}\left(\left|\Psi_{BC}^{(+)}\right\rangle - \left|\Psi_{BC}^{(-)}\right\rangle\right),$$

$$|1_B\rangle|1_C\rangle = \frac{1}{\sqrt{2}}\left(\left|\Phi_{BC}^{(+)}\right\rangle - \left|\Phi_{BC}^{(-)}\right\rangle\right).$$

If we reorder the qubits, putting forward the subsystems A and D, after some calculations the initial four-qubit state can be rewritten as:

$$|\psi_{ADBC}\rangle = \frac{1}{2}\bigg[\frac{1}{\sqrt{2}}(|0_A\rangle|0_D\rangle + |1_A\rangle|1_D\rangle)\left|\Phi_{BC}^{(+)}\right\rangle$$

$$+ \frac{1}{\sqrt{2}}(|0_A\rangle|0_D\rangle - |1_A\rangle|1_D\rangle)\left|\Phi_{BC}^{(-)}\right\rangle$$

$$+ \frac{1}{\sqrt{2}}(|0_A\rangle|1_D\rangle + |1_A\rangle|0_D\rangle)\left|\Psi_{BC}^{(+)}\right\rangle$$

$$+ \frac{1}{\sqrt{2}}(|0_A\rangle|1_D\rangle - |1_A\rangle|0_D\rangle)\left|\Psi_{BC}^{(-)}\right\rangle\bigg],$$

or, exploiting the Pauli matrices:

$$|\psi_{ADBC}\rangle = \frac{1}{2}\bigg[|\psi_{AD}\rangle\big|\Phi_{BC}^{(+)}\big\rangle + \hat{\mathbb{I}} \otimes \hat{\sigma}_z^{(D)}|\psi_{AD}\rangle\big|\Phi_{BC}^{(-)}\big\rangle$$

$$+ \hat{\mathbb{I}} \otimes \hat{\sigma}_x^{(D)}|\psi_{AD}\rangle\big|\Psi_{BC}^{(+)}\big\rangle + \hat{\mathbb{I}} \otimes \big(\hat{\sigma}_x^{(D)}\hat{\sigma}_z^{(D)}\big)|\psi_{AD}\rangle\big|\Psi_{BC}^{(-)}\big\rangle\bigg],$$

where:

$$|\psi_{AD}\rangle = \frac{|0_A\rangle|0_D\rangle + |1_A\rangle|1_D\rangle}{\sqrt{2}}.$$

It is now clear that, as in the case of the "standard" teleportation protocol, after performing the Bell measurement and applying the suitable unitary operation according to the measurement outcome, the state of the qubits A and D is $|\psi_{AD}\rangle$.

3.2 We recall that the standard computational process requires that:

$$\hat{U}_f|x\rangle|0\rangle = |x\rangle|f(x)\rangle.$$

Thereafter, it is easy to verify that the following unitary operators implement the action of the functions listed in the problem:

(a) $f(0) = f(1) = 0 \;\Rightarrow\; \hat{U}_f = \hat{\mathbb{I}} \otimes \hat{\mathbb{I}}$;
(b) $f(0) = f(1) = 1 \;\Rightarrow\; \hat{U}_f = \hat{\mathbb{I}} \otimes \hat{\sigma}_x$;
(c) $f(0) \neq f(1) = 0 \;\Rightarrow\; \hat{U}_f = \big(\hat{\mathbb{I}} \otimes \hat{\sigma}_x\big)$ CNOT;
(d) $f(0) \neq f(1) = 1 \;\Rightarrow\; \hat{U}_f =$ CNOT.

We remark that these implementations are not unique.

The interested reader can also write quantum circuit associated with the unitary operators and study what happens to the operator $\mathbf{H} \otimes \mathbf{H}\hat{U}_f\mathbf{H} \otimes \mathbf{H}$, that is the quantum solution of the Deutsch problem.

3.3 If $a = 0$ it is clear that $f(x) = 0$, $\forall x \in \{0, 1, \ldots, 2^n - 1\}$. Now we focus on the case $a \neq 0$ and we prove that, in this case, the function is balanced, that is $f(x) = 0$ for half of the possible 2^n input values and $f(x) = 1$ for the other half.

As in the case of the Deutsch–Jozsa problem, after the application of the Hadamard transformations the final state of the input register can be written as:

$$|\psi\rangle_n = \sum_{z=0}^{2^n-1} c_f(z)|z\rangle_n,$$

with

$$c_f(z) = \frac{1}{2^n} \sum_{x=0}^{2^n-1} (-1)^{z \cdot x + f(x)}.$$

Due to the action of the unitary transformation implementing $f(x)$, that is based on CNOT gates as sketched in Fig. 3.19, if the initial states of the input and output registers are $|0\rangle_n$ and $|1\rangle$, respectively, then the final state of the output register is $|a\rangle_n \neq |0\rangle_n$ when $a \neq 0$ (see the bottom panel of Fig. 3.20). Thereafter, we conclude that one must have:

$$c_f(0) = \frac{1}{2^n} \sum_{x=0}^{2^n-1} (-1)^{f(x)} = 0,$$

but this is true if and only if $f(x) = 0$ for half of the possible 2^n input values and $f(x) = 1$ for the other half, namely, the function is balanced.

Problems of Chap. 5

5.1 Given $\hat{T}$, it is clear that the two eigenvectors are $|0\rangle$ and $|1\rangle$ with eigenvalues:

$$\hat{T}|0\rangle = |0\rangle \quad \text{and} \quad \hat{T}|1\rangle = e^{2\pi i \phi}|0\rangle,$$

with $\phi = 1/8 = 0.125$. Therefore, the input four-qubit state of the protocol is $|0\rangle_3 \otimes |1\rangle$.

After the application of the Hadamard transformations to the input register we have:

$$\mathbf{H}^{\otimes 3} \otimes \hat{\mathbb{I}} |0\rangle_3 \otimes |1\rangle = \frac{1}{2^{3/2}} \sum_{x=0}^{2^3-1} |x\rangle_3 \otimes |1\rangle$$

$$= \bigotimes_{k=1}^{3} \frac{|0_k\rangle + |1_k\rangle}{\sqrt{2}} \otimes |1\rangle,$$

where the integer x has the binary expansion $x = x_1 \, 2^2 + x_2 \, 2 + x_3 \, 2^0$, and we write $|x\rangle_3 = |x_1\rangle|x_2\rangle|x_3\rangle$.

Now, we apply the conditional gates. Since:

$$\hat{T}_k^{2^{3-k}} \frac{|0_k\rangle + |1_k\rangle}{\sqrt{2}} \otimes |1\rangle = \frac{|0_k\rangle + \exp\left(2\pi i \phi \, 2^{3-k}\right)|1_k\rangle}{\sqrt{2}} \otimes |1\rangle,$$

we get (due to the definition of $|x\rangle_3$, we have also to take care of the conditional gates, which have a different control qubit with respect to Fig. 5.4):

$$\hat{T}_1^4 \hat{T}_2^2 \hat{T}_3 \frac{1}{2^{3/2}} \sum_{x=0}^{2^3-1} |x\rangle_3 \otimes |1\rangle = \frac{1}{2^{3/2}} \sum_{x=0}^{2^3-1} e^{2\pi i \phi x} |x\rangle_3 \otimes |1\rangle.$$

As final step we use of the inverse of the quantum Fourier transform onto the three-qubit state:

$$\hat{F}_Q^{-1}\left(\frac{1}{2^{3/2}} \sum_{x=0}^{2^3-1} e^{2\pi i \phi x} |x\rangle_3\right) = \frac{1}{2^3} \sum_{x=0}^{2^3-1} e^{2\pi i \phi x} \sum_{y=0}^{2^3-1} \exp\left(-2\pi i \phi \frac{y x}{2^3}\right) |y\rangle_3,$$

$$= \frac{1}{2^3} \sum_{y=0}^{2^3-1} \sum_{x=0}^{2^3-1} \exp\left(-2\pi i \phi x \frac{y - \phi 2^3}{2^3}\right) |y\rangle_3,$$

$$\underbrace{\phantom{\sum_{y=0}^{2^3-1} \sum_{x=0}^{2^3-1}}}_{2^3 \delta_{0,(y-\phi 2^3)}}$$

$$= \left|\phi 2^3\right\rangle = |0_1\rangle|0_2\rangle|1_3\rangle \equiv |\varphi_1\rangle|\varphi_2\rangle|\varphi_3\rangle,$$

where we explicitly put $\phi = 1/8$ (we are forced to use this information to carry on the calculations, but the quantum algorithm just use the action of the $\hat{T}$ gate "without knowing" the actual value of ϕ) and:

$$\phi 2^3 = 2^3 \sum_{m=1}^{3} \varphi_m 2^{-m} = \sum_{k=0}^{2} \varphi_{k+1} 2^k.$$

Finally, being $\phi = 0.\varphi_1\varphi_2\varphi_3$ (binary expansion), we find $\phi = 0.125$.

Problems of Chap. 6

6.3 The Hamiltonian of the system is (we drop, for simplicity, the subscript n):

$$\hat{H} = -\gamma A - \sum_{w \in B} |w\rangle\langle w|,$$

A being the adjacency matrix, and we want to calculate the matrix elements:

$$\langle \beta|\hat{H}|\beta\rangle, \quad \langle \alpha|\hat{H}|\alpha\rangle \quad \text{and} \quad \langle \beta|\hat{H}|\alpha\rangle.$$

where:

$$|\alpha\rangle = \frac{1}{\sqrt{N-M}} \sum_{x \in A} |x\rangle, \quad \text{and} \quad |\beta\rangle = \frac{1}{\sqrt{M}} \sum_{w \in B} |w\rangle,$$

We obtain:

$$\langle\beta|\hat{H}|\beta\rangle = -\gamma \langle\beta|A|\beta\rangle - \langle\beta| \underbrace{\left(\sum_{w \in B} |w\rangle\langle w|\right)}_{1} |\beta\rangle,$$

with

$$\langle\beta|A|\beta\rangle = \frac{1}{M} M(M-1) = M - 1,$$

since each solution vertex is connected with the other $M-1$ solution vertices. Analogously, we find:

$$\langle\beta|\hat{H}|\beta\rangle = -\gamma \langle\alpha|A|\alpha\rangle - \langle\alpha| \underbrace{\left(\sum_{w \in B} |w\rangle\langle w|\right)}_{0} |\alpha\rangle,$$

with:

$$\langle\alpha|A|\alpha\rangle = \frac{1}{N-M} \sum_{x \in A} \sum_{y \in A} \langle x|A|y\rangle,$$

$$= \frac{1}{N-M} (N-M)(N-M-1) = N - M - 1,$$

because each of the $N-M$ not-solution vertex is linked to the other $N-M-1$ not-solution vertices.

The off-diagonal element reads (note that $\langle\beta|\hat{H}|\alpha\rangle = \langle\alpha|\hat{H}|\beta\rangle$):

$$\langle\beta|\hat{H}|\alpha\rangle = -\gamma \langle\beta|A|\alpha\rangle - \langle\beta| \underbrace{\left(\sum_{w \in B} |w\rangle\langle w|\right)}_{0} |\alpha\rangle,$$

with:

$$\langle \beta | A | \alpha \rangle = \frac{1}{\sqrt{M(N-M)}} \sum_{w \in B} \sum_{x \in A} \langle w | A | x \rangle,$$

$$= \frac{1}{\sqrt{M(N-M)}} M(N-M) = \sqrt{M(N-M)},$$

each solution vertex being connected with the $N - M$ not-solution vertices. Eventually, we can write:

$$\hat{H} = -\gamma \begin{pmatrix} M - 1 + \gamma^{-1} & \sqrt{M(N-M)} \\ \sqrt{M(N-M)} & N - M - 1 \end{pmatrix}.$$

Problems of Chap. 7

7.1 The 4×4 matrix associated with the state $\hat{\varrho}_{AB}$ (we use the computational basis) reads:

$$\hat{\varrho}_{AB} = \frac{q}{2} \begin{pmatrix} 0 & 0 & 0 & 0 \\ 0 & 1 & -1 & 0 \\ 0 & 1 & 1 & 0 \\ 0 & 0 & 0 & 0 \end{pmatrix} + \frac{1-q}{4} \begin{pmatrix} 1 & 0 & 0 & 0 \\ 0 & 1 & 0 & 0 \\ 0 & 0 & 1 & 0 \\ 0 & 0 & 0 & 1 \end{pmatrix}$$

and, writing q as a function of p, we obtain:

$$\hat{\varrho}_{AB} = \begin{pmatrix} \frac{1}{3}p & 0 & 0 & 0 \\ 0 & \frac{1}{6}(3-2p) & -\frac{1}{6}(3-4p) & 0 \\ 0 & -\frac{1}{6}(3-4p) & \frac{1}{6}(3-2p) & 0 \\ 0 & 0 & 0 & \frac{1}{3}p \end{pmatrix}.$$

Therefore, the partially transposed matrix is:

$$\mathcal{T}_A \otimes \hat{\mathbb{I}}_B (\hat{\varrho}_{AB}) = \begin{pmatrix} \frac{1}{3}p & 0 & 0 & -\frac{1}{6}(3-4p) \\ 0 & \frac{1}{6}(3-2p) & 0 & 0 \\ 0 & 0 & \frac{1}{6}(3-2p) & 0 \\ -\frac{1}{6}(3-4p) & 0 & 0 & \frac{1}{3}p \end{pmatrix}.$$

whose eigenvalues are:

$$\tilde{\lambda}_1 = p - \frac{1}{2}, \quad \tilde{\lambda}_2 = \tilde{\lambda}_3 = \tilde{\lambda}_4 = \frac{1}{2}\left(1 - \frac{2}{3}\right) > 0.$$

Therefore, if $p < 1/2$ or, equivalently, $q > 1/3$, one finds $\tilde{\lambda}_1 < 0$ and the partially transposed state is no longer positive.

To answer to the question about the entanglement of $\hat{\varrho}_{AB}$, we can calculate its concurrence. It is straightforward to find the eigenvectors λ_k and eigenvalues $|\lambda_k\rangle$ of $\hat{\varrho}_{AB}$, namely:

$$\lambda_1 = 1 - p, \quad |\lambda_1\rangle = |\Psi_{AB}\rangle = \frac{|0_A\rangle|1_B\rangle - |1_A\rangle|0_B\rangle}{\sqrt{2}},$$

$$\lambda_2 = \frac{p}{3}, \quad |\lambda_2\rangle = \frac{|0_A\rangle|1_B\rangle + |1_A\rangle|0_B\rangle}{\sqrt{2}},$$

$$\lambda_3 = \frac{p}{3}, \quad |\lambda_3\rangle = |0_A\rangle|0_B\rangle,$$

$$\lambda_4 = \frac{p}{3}, \quad |\lambda_4\rangle = |1_A\rangle|1_B\rangle.$$

It is easy to verify that the state $\hat{\varrho}'_{AB} = \hat{\sigma}_y \otimes \hat{\sigma}_y \hat{\varrho}^*_{AB} \hat{\sigma}_y \otimes \hat{\sigma}_y$, we need to calculate the concurrence, has the same eigenvectors and eigenvalues of $\hat{\varrho}_{AB}$ and, thus, we simply have (see Sect. 2.8):

$$\hat{R} = \sqrt{\sqrt{\hat{\varrho}_{AB}}\, \hat{\varrho}'_{AB} \sqrt{\hat{\varrho}_{AB}}} = \hat{\varrho}_{AB},$$

and the concurrence reduces to:

$$C(\hat{\varrho}_{AB}) = \max(0, \lambda_1 - \lambda_2 - \lambda_3 - \lambda_4),$$

$$= \max(0, 1 - 2p),$$

therefore, the state is entangled if $p < 1/2$ (or $q > 1/3$).

We also conclude that $\hat{\varrho}_{AB}$ is entangled if and only if it is no longer positive semi-definite under the partial transposition operation. It is worth noting that the positivity of the partial transpose (PPT criterion) is a *necessary* condition for the separability that is also *sufficient* in the case of two-qubits, like the present one, or qubit-qutrit systems.[1]

---

[1] A. Peres, *Separability Criterion for Density Matrices*, Phys. Rev. Lett. **77**, 1413–1415 (1996); M. Horodecki, P. Horodecki, and R. Horodecki, *Separability of Mixed States: Necessary and Sufficient Conditions*, Phys. Lett. A **223**, 1–8 (1996).

Problems of Chap. 11

11.1 If we apply the operator:

$$\hat{O} = \frac{1}{2} \sum_{N=-\infty}^{+\infty} (|N\rangle\langle N+1| + |N+1\rangle\langle N|).$$

to the state:

$$|\varphi\rangle = \frac{1}{\sqrt{2\pi}} \sum_{N=-\infty}^{+\infty} e^{iN\varphi} |N\rangle$$

we get:

$$\hat{O}|\varphi\rangle = \frac{1}{\sqrt{2\pi}} \sum_{N=-\infty}^{+\infty} \sum_{M=-\infty}^{+\infty} e^{iM\varphi} \left(|N\rangle\langle N+1| + |N+1\rangle\langle N|\right)|M\rangle,$$

$$= \frac{1}{2\sqrt{2\pi}} \left[\sum_{N=-\infty}^{+\infty} e^{i(N+1)\varphi} |N\rangle + \sum_{N=-\infty}^{+\infty} e^{iN\varphi} |N+1\rangle \right],$$

$$= \frac{1}{2\sqrt{2\pi}} \left[e^{i\varphi} \sum_{N=-\infty}^{+\infty} e^{iN\varphi} |N\rangle + e^{-i\varphi} \sum_{N=-\infty}^{+\infty} e^{iN\varphi} |N\rangle \right],$$

$$= \cos\varphi \, |\varphi\rangle.$$

Therefore, $|\varphi\rangle$ is an eigenvector of $\hat{O}$ with eigenvalue $\cos\varphi$. Since:

$$\cos\hat{\varphi} \, |\varphi\rangle = \cos\varphi \, |\varphi\rangle,$$

the two operators have the same spectral decomposition and we can conclude that:

$$\cos\hat{\varphi} = \frac{1}{2} \sum_{N=-\infty}^{+\infty} (|N\rangle\langle N+1| + |N+1\rangle\langle N|).$$

Index

A
Adiabatic
  quantum computation, 183
  theorem, 185
Algorithm
  Bernstein–Vazirani, 47
  continued-fraction, 84
  Deutsch, 43
  Deutsch–Jozsa, 45
  factoring, 79, 85
  Grover, 91
  order-finding, 81
  phase estimation, 74
  RSA, 88
  search, 91
  Shor, 79, 85, 88
Amplitude damping channel, 114
  generalized, 116

B
Bell
  measurement, 37, 39
  states, 37, 38
Binary symmetric channel, 120
Bit flip, 111
Bit-phase flip, 112
Bloch
  sphere, 14, 33
  vector, 14, 22
Born rule, 16
Boson sampling, 145

C
Cavity QED, 141
Choi–Jamiołkowski isomorphism, 106
Church–Turing thesis, 64

Circuit identities, 41
Clifford
  gates, 60
  group, 60
Coherent states, 146
Completely positive map, 105, 106, 109, 119
Complexity classes
  MA, 66, 67
  NP, 66, 67
  NP-complete, 66
  NP-hard, 66
  P, 65, 67
  PSPACE, 65
Concurrence, 27
Conditional states, 24
Controlled not (CNOT), 6, 8, 34
  Cirac–Zoller, 161
  with spin systems, 135, 136
  with trapped ions, 161
Controlled unitary, 35
Cooper pair box (CPB), 172
  Hamiltonian, 173
  two-level Hamiltonian, 175
Cryptography
  asymmetric, 88
  RSA, 88

D
Decision problems, 65
Density
  matrix, 20
  operator, 20
    single qubit, 22
Depolarizing channel, 113
Dressed states, 148

Index

E
Entanglement, 25
  concurrence, 27
  entropy of, 25
  measure of, 25
  swapping, 61
Entropy
  of entanglement, 25
  von Neumann, 25
Error correction
  parity-check, 123
  three-bit code, 120
Error syndrome, 119
Euler
  theorem, 80
  totient function, 80, 88
Exchange interaction, 136

F
Fabry–Perot cavity, 197
Factorization, 79, 85, 88
  through adiabatic evolution, 191
Fault-tolerant quantum computation, 33, 127, 146
Fock state, 145
Fredkin gate, 51

G
Gaussian states, 145
GKP code, 146
Gottesman–Knill theorem, 60
Graph, 99
Gray code, 55
Grover iterator, 92

H
Hadamard gate, 32, 42
  and binary function, 43
  with $\frac{1}{2}$-spin, 133

I
Interaction picture, 195
Ion crystal, 156
Ising interaction, 135

J
Jaynes–Cummings
  Hamiltonian, 147, 148
  model, 146, 162

dressed states, 148
vacuum Rabi oscillations, 149
Josephson
  energy, 180
  equations, 167
  junction, 166
    classical Hamiltonian, 168
    quantum Hamiltonian, 170
  plasma frequency, 179

K
KLM protocol, 145
Kraus operator, 104

L
Lamb–Dicke
  parameter, 158
  regime, 159
Larmor frequency, 132

N
No-cloning theorem, 36
Nuclear magnetic resonance (NMR), 133, 140

O
Operation element, 104
Operator sum representation, 104
Orbital angular momentum of light, 145
Order-finding protocol, 81

P
Parity check, 123
Partial trace, 22
Partial transposition, 105
  and entanglement, 118
  and Werner states, 118
Pauli matrices, 8
Paul trap, 153
Phase damping channel, 117
Phase estimation protocol, 74
Phase flip, 112, 118
Phase gate, 33
Phase kickback, 43, 44, 46, 74
Phase shift gate, 33
Positive operator-valued measure (POVM), 28
Problem
  halting, 66
  travelling salesman, 66
  unsolvable, 67

Index

Purification, 24
Purity, 21

Q

Quantum complexity classes, 65
   BQP, 67
   PSPACE, 65
   QMA, 67
Quantum error correction
   conditions, 120
   error syndrome, 119
   recovery, 120
   Shor code, 126
   syndrome measurement, 119
   three-qubit code, 121
Quantum Fourier transform, 69
   quantum circuit, 73
Quantum mechanics
   postulates, 15
Quantum operation, 103, 120
   bit flip, 111
   bit-phase flip, 112
   operator sum representation, 104
   phase flip, 112
   physical interpretation, 106
Quantum teleportation, 38
Quantum walks, 99
   adjacency matrix, 99
   Laplacian, 99
Qubit
   charge, 171, 172
      with transmission line, 175
   definition, 13
   hyperfine, 164
   multiple qubit state, 15
   optical, 164
   path, 145
   photonic, 145
   polarization, 145
   spin, 132
   superconducting, 165
   transmon, 171, 177, 180
   trapped ion, 160
   two-level atom, 141

R

Rabi frequency, 143, 159
Rivest Shamir Adleman (RSA), 88
Rotating-wave approximation, 134, 142

S

Spin
   electronic, 140
   Larmor precession, 132
   nuclear, 140
   spin–1/2, 131
Squeezed states, 145
Stabilizer codes, 127
Standard computational process, 40
Stean code, 128
Superconducting QUantum Interference Device, 166, 169
SWAP, 5, 8, 35
Syndrome measurement, 119

T

Thermal equilibrium, 116
Threshold theorem, 128
Toffoli gate, 49, 124
Trotter formula, 98
Turing machine, 63
   probabilistic, 64
   quantum, 64
   universal, 64
Two-level system, 16, 29
   atom, 141
   charge qubit, 175
   spin–1/2, 131
   trapped ion, 160

U

Universal quantum computation, 145
Universal quantum gates, 52
   set of, 57, 127

V

Vacuum Rabi oscillations, 149
Von Neumann entropy, 25

W

Werner states, 118